D0961452

DARWIN COMES TO TOWN

MENNO SCHILTHUIZEN

DARWIN COMES TO TOWN

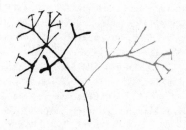

HOW THE URBAN JUNGLE DRIVES EVOLUTION

PICADOR | NEW YORK

picadorusa.com • instagram.com/picador
twitter.com/picadorusa • facebook.com/picadorusa

Picador® is a U.S. registered trademark and is used by Macmillan Publishing Group, LLC, under license from Pan Books Limited.

For book club information, please visit facebook.com/picadorbookclub or email marketing@picadorusa.com.

Photographic acknowledgements: chapter 1, AntRoom/Taku Shimada; chapter 2, Sow-Yan Chan; chapter 3, Menno Schilthuizen; chapter 4, Natural History Museum Rotterdam; chapter 5, Getty Images; chapter 6, Herman Berkhoudt; chapter 7, Wikimedia Commons; chapter 8, Menno Schilthuizen; chapter 9, Stephanie Doucet; chapter 10, Ariane Le Gros; chapter 11, Marion Chatelain; chapter 12, FLAP; chapter 13, Axel Hochkirch; chapter 14, Kim Meijer; chapter 15, Getty Images; chapter 16, Millie Mockford; chapter 17, David Rentz; chapter 18, Luis De León; chapter 19, Natural History Museum Rotterdam; chapter 20, Fenna Schilthuizen

Designed by Jonathan Bennett

The Library of Congress Cataloging-in-Publication Data is available upon request.

ISBN 978-1-250-12782-2 (hardcover)
ISBN 978-1-250-12783-9 (ebook)

Our books may be purchased for educational, business, or promotional use. For information on bulk purchases, please contact the Macmillan Corporate and Premium Sales Department at 1-800-221-7945, extension 5442, or write to specialmarkets@macmillan.com.

Originally published in Great Britain by Quercus Books

First U.S. Edition: April 2018

10 9 8 7 6 5 4 3 2 1

for Iva

CONTENTS

III. CITY ENCOUNTERS

IV. DARWIN CITY

DARWIN COMES TO TOWN

CITY PORTAL

IT'S PERFECTLY FORMED. A MIRACLE of micro-engineering ready for its short visit to the world. Gossamer wings, still unfrayed, folded carefully over its imperceptibly breathing abdomen. Six nimble legs, delicately placed on the dusty wall, in mint condition—each with a complete set of nine segments, not yet diminished in number by sharp connections with ventilation fans or jumping spiders' forelegs. Its golden-bristled thorax, a power nugget encasing the balled-up energy of flight muscles, so massive that it almost hides from view the serene face behind which a miniaturized brain coordinates input and output channels of antennae, palps, all-seeing eyes, and the eight interlocking sheaths of its parasitic proboscis.

I am standing in a hot and crowded pedestrian tunnel of London Underground's Liverpool Street station, spectacles in one hand, nose pressed against the tiled wall, admiring this fine specimen of the house mosquito, *Culex molestus*, freshly emerged from its pupa. But I am slowly waking up from my entomological reverie. Not only because of the rushed passersby that avoid knocking into me with a last-second swerve and a muttered "excuse me" that is more accusatory than apologetic, but also because I am becoming uncomfortably aware of the CCTV camera suspended from the ceiling and Transport for London's repeatedly broadcast advice to its passengers to report any suspicious behavior to a member of staff.

For a biologist, the inner city is an unlikely place for any professional activities. The unwritten rule among biologists is that, when prompted, one should answer gruffly that cities are only necessary evils where a true biologist spends as little time as possible. The real world lies outside the urban realm, in forests, dales, and fields. Where the wild things are.

But if I am honest I must admit I secretly like cities. Not so much the organized, slick, well-oiled parts of them, but rather the grimy, organic fabric of the city, revealed in forgotten corners where the threadbare carpet of culture gives way—the city's underbelly where the artificial and the natural meet and engage in ecological relations. To my biologist's eye, the inner city, for all its hustle and bustle and thoroughly unnatural appearance, becomes a constellation of miniature ecosystems. Even in these seemingly sterile, thoroughly brick-and-concrete-clad streets of Bishopsgate, I spot life-forms that cling on in stubborn defiance. Here, a snapdragon growing in wild profusion from some invisible crack in the plastered wall of a flyover. There, the unspeakable chemistry between cement and leaked sewage from which glassy off-white icicles are born, which in turn serve as anchor points for the soot-dusted webs of common orb-weaver spiders. The emerald veins of moss sprouting from the slits between cracked reinforced glass and its frame, fighting for supremacy with the rust bubbles working their way through the red lead paint. Feral rock pigeons with diseased feet balance among the plastic spines on a ledge. (Somebody has put up a sticker next to it showing an enraged pigeon with balled wing-fists, saying: "Plastic needles represent a cynical, oppressive restriction of our right to free assembly. This fight is not over!") And a mosquito on the wall of a train station underpass.

It is not just any mosquito. *Culex molestus* is also known as the London Underground mosquito. It gained this name first because of the havoc it wreaked on Londoners sheltering on the platforms and tracks of the Central line at Liverpool Street station during the German bombing raids of the city in 1940. And then, in the 1990s, because of the interest taken in them by University of London geneticist Katharine Byrne. Byrne joined maintenance crews on their daily expeditions into the bowels of the city's tube network. They went into the deepest parts of the tunnels where the brick walls supporting tangles of wrist-thick electric cables are blackened by dust from the trains' brake shoes, and where the only place indicators are mysterious codes in chalk, spray paint, and ancient enameled plates. Here, the London Underground mosquitoes live and breed. They steal the blood of commuters, and lay their eggs in flooded sumps and shafts, which is where Byrne collected their larvae.

She took samples of water-with-larvae from seven places on the Central, Victoria, and Bakerloo lines, brought them to her lab, waited for the larvae to develop into adult mosquitoes (like the one I saw on that tunnel wall) and then extracted their proteins for genetic analysis. Twenty years ago, I saw her present her results at a conference in Edinburgh. Even though her audience consisted of seasoned evolutionary biologists, she managed to thrill us all. First, the underground mosquitoes in the three tube lines were genetically different from one another. This was, Byrne told us, because the tube lines form nearly separate worlds, with the clouds of mosquitoes in each line stirred and mixed by the constant piston-like action of trains moving around in snugly-fitting tunnels. The only way for the mosquitoes in the Central, Bakerloo and Victoria lines to become genetically mixed, she pointed out, would be "for all of them to change trains at Oxford Circus station." But not only were the mosquitoes in separate underground

tubes different from each other. They were also different from their
above-ground relatives. Not just in their proteins, but also in their
way of life. Up on London's streets, the mosquitoes feed on bird,
not human, blood. They need a blood meal before they can lay their
eggs, they mate in large swarms, and they hibernate. Down in the
tube, the mosquitoes suck commuters' blood and lay eggs before
feeding; they don't form mating swarms but seek their sexual pleasures in confined spaces, and are active the whole year round.

Since Byrne's work, it has become clear that the Underground
mosquito is not unique to London. It lives in cellars, basements
and subways all over the world, and it has adapted its ways to its
human-sculpted environment. Thanks to mosquitoes that get
trapped in cars and planes, its genes spread from city to city,
but at the same time it also cross-breeds with local above-ground
mosquitoes, absorbing genes from that source as well. And it has
also become clear that all this has happened very, very recently—
probably only since humans began constructing underground
buildings, did *Culex molestus* evolve.

As I take a last good look at my very own London Underground
mosquito, in that crowded passageway in Liverpool Street station, I imagine the invisible modifications that evolution has ac-
complished inside that tiny, fragile body. Proteins in its antennae
have changed shape so that our human odors, rather than bird
smells, elicit a response. Genes that regulate its biological clock
have been reset or turned off, to prevent it from going into hiber-
nation, since there is always human blood underground and it
never gets very cold. And think of the complex diversifications
that have been needed for the change in sexual behavior! From a
species where the males swarm in large clouds that females dart in
and out of to seek fertilization, to one that mates by simple one-
on-one pairing in the small spaces where the sparsely distributed
underground mosquitoes happen to run into each other.

The evolution of the London Underground mosquito speaks to our collective imagination. Why are we so tickled by it and why do I so vividly remember Katharine Byrne's presentation from all those years ago? First, we have been taught that evolution is a slow process, imperceptibly whittling species over millions of years—not something that could take place within the short timespan of human urban history. It drives home the fact that evolution is not only the stuff of dinosaurs and geological epochs. It can actually be observed here and now! Secondly, the notion that our impact on the environment is so great that "wild" animals and plants are actually adapting to habitats that were originally created by humans for humans, makes us aware that some of the changes we are enforcing on the earth are irreversible.

The third reason why we prick up our ears when hearing about the London Underground mosquito is because it seems such a cute addition to evolution's standard portfolio. We all know about evolution perfecting the plumage of birds of paradise in faraway jungles or the shape of orchid flowers on lofty mountaintops. But apparently, the process is so mundane that it is not above plying its trade below our feet, among the grimy power cables of the city's metro system. Such a nice, unique, close-to-home example! The sort of thing you'd expect to find in a biology textbook.

But what if it is not an exception anymore? What if the Underground mosquito is representative of all flora and fauna that come into contact with humans and the human-crafted environment? What if our grip on the earth's ecosystems has become so firm that life on earth is in the process of evolving ways to adapt to a thoroughly urban planet? These are the questions we will be tackling in this book.

None too soon, either. In 2007, the world passed a crucial benchmark: in that year, for the first time in history, the number of people living in urban areas outnumbered those living in rural areas. Since then, that statistic has been rising rapidly. By mid-twenty-first century, two-thirds of the world's estimated 9.3 billion will be in cities. Mind you: that's for the entire world. In western Europe, more people have lived in cities than in the countryside since 1870, and in the US that tipping point was reached in 1915. Areas like Europe and North America have been firmly on the way to becoming urban continents for more than a century. A recent study in the US showed that each year, the average distance between a given point on the map and the nearest forest, increases by about 1.5 percent.

Never before in the history of our planet has a single life form been so dominant. "Well, what about the dinosaurs?" you might ask. But the dinosaurs were an entire class of animals, probably thousands of species. Comparing the thousands of species of dinosaurs with the single species of *Homo sapiens* would be like comparing all the world's sole-proprietor greengrocers with Tesco's. No, in ecological terms the world has never before seen the situation that we find ourselves in today: a single large animal species completely and utterly blanketing the planet and turning it to its advantage. At the moment, our species appropriates fully one-quarter of the food that all of the world's plants produce and half of all the world's freshwater run-off. Again, something that has never happened before: no other species that evolution has produced has ever been able to play such a central ecological role on such a global scale.

So, our world is becoming thoroughly human-dominated. By 2030, nearly 10 percent of the landmass of the planet will be urbanized, and much of the rest covered by human-shaped farms, pasture, and plantations. Altogether a set of entirely new habitats, the likes of which nature has not seen before. And yet, when we talk about ecology and evolution, about ecosystems and nature, we are stubbornly factoring out humans, myopically focusing our attention on that diminishing fraction of habitats where human influence is still negligible. Either that, or we are trying to quarantine nature, as much as possible, from the harmful impacts of the human, implicitly non-natural, world.

Such an attitude can no longer be maintained. It's time to own up to the fact that human actions are the world's single most influential ecological force. Whether we like it or not, we have become fully integrated with everything that goes on on this planet. Only in our flights of fancy can we still keep nature divorced from the human environment. Out in the real world, our tentacles firmly entwine nature's fabric. We build cities full of novel structures made of glass and steel. We irrigate, pollute, and dam waterways; mow, spray, and fertilize fields. We pump greenhouse gases into the air that alter the climate; we release non-native plants and animals, and harvest fish, game, and trees for our food and other needs. Every non-human life form on earth will come across humans, either directly or indirectly. And, mostly, such encounters are not inconsequential for the organism in question. They may threaten its survival and way of life. But they may also create new opportunities, new niches. Like they did for the ancestors of *Culex molestus*.

So what does nature do when it meets challenges and opportunities? It evolves. If at all possible, it changes and adapts. The greater the pressure, the faster and more pervasive it does so. As the neck-tied traders who rush past me in that Liverpool Street station tunnel know all too well, in cities there is great opportunity, but also great competition. Every second counts if you want to survive. In this book, I will show that nature is doing just that. While we all have been focusing on the vanishing quantity of unspoiled nature, urban ecosystems have been evolving behind our backs, right in the cities that we have been turning up our naturalist noses at. While we have been trying to save the world's crumbling pre-urban ecosystem, we have been ignoring the fact that nature has already been putting up the scaffolds to build novel, urban ecosystems for the future.

I will reveal the myriad ways in which urban ecosystems are assembling themselves and how they might, one day, be the chief form of nature on our urbanized planet. But before we take off, there is something I need to get off my chest.

The growing band of people who try to generate an appreciation for nature in the urban environment often get accused of providing excuses for developers to destroy wild nature—or even of getting into bed with the enemy, and stabbing nature conservation in the back. Several years ago, with my colleague Jef Huisman of the University of Amsterdam, I wrote an opinion article in the Dutch national newspaper *De Volkskrant*, in which we argued that nature is dynamic, constantly in flux, and we should not try to preserve Dutch ecosystems in exactly the same shape and composition as we see them in landscape paintings of several centuries ago. We argued for a more pragmatic approach to conservation in which there is a place for exotic species, urban nature, and more attention to the smooth running of the ecosystem, rather than to the exact species therein.

That did not go down too well with some people. We received angry emails from colleagues who accused us of playing into the hands of right-wing politicians who would seize on the flimsiest of excuses to continue their rampage over the natural world. Other irate readers advised us to "tell that to the people of Australia and New Zealand, who see their nature overrun by cane-toads and rabbits."

Such attacks hurt me deeply. I grew up as a bug-collecting, bird-watching boy, spending days on end by myself in the fields surrounding my hometown, armed with a pair of binoculars, a plant guide, or a jar for collecting beetles. Today, the fields where

I photographed nesting godwits, stepped through carpets of early marsh-orchids, and caught my first great silver water beetle, have been absorbed by the urban sprawl of the Rotterdam conurbation. I looked on in furious, tearful impotence, with balled fists, as the first bulldozers began leveling my playground and swore to avenge the nature that had been lost forever. Later, as a tropical ecologist living and working in Borneo, I watched powerlessly as mangroves were converted into parking lots and pristine rainforests to oil palm monocultures.

But that same love and care for nature also gave me an understanding of the power of evolution and the relentless adaptability of the living world. The expansion of the human population is a given. Barring global disaster or dictatorial birth control, humans will be smothering the earth with their cities and urbanized environments before the century is up. For that reason, we must conserve as many unspoiled wilderness areas as possible, and this book should not be misconstrued to devalue such efforts. However, at the same time, we must realize that outside of pristine areas, traditional conservation practices (eradicating exotic species, vilifying "weeds" and "pests") may in fact be destroying the very ecosystems that are going to sustain humankind in the future. Instead, I argue in this book that we must embrace and harness the evolutionary forces that are shaping novel ecosystems right here, right now, and work toward allowing nature to grow in the hearts of our cities.

I.

CITY LIFE

*Numberless crowded streets, high growths of iron, slender,
strong, light, splendidly uprising toward clear skies*

WALT WHITMAN, "MANNAHATTA" (*Leaves of Grass*, 1855)

1
NATURE'S ULTIMATE
ECOSYSTEM ENGINEER

SOME 20 MILES WEST OF THE CITY OF
Rotterdam lie the coastal sand dunes of Voorne—an extensive area (at least, by Dutch diminutive standards) of rolling,
vegetated dunes, though increasingly consumed from the north
by Rotterdam's expanding port. You can sit there, with your
buttocks on a carpet of mosses and lichens, eating a sandwich
among the rare yellow-wort and marsh helleborines, while
in the distance gigantic heaps of iron ore and coal are shifted
around, the cling and clang wafting in and out on the incessant
wind.

It is here that I spent almost every Saturday as a schoolboy,
hunting beetles for my expanding collection. My juvenile-naturalist

friends and I, sometimes accompanied by our indefatigable biology teacher, would cycle along the Meuse, take the ferry across the river, zigzag among the oil-storage tanks and daunting chemical installations of the refineries, and then spend a whole day in the dunes, botanizing and entomologizing. Sundays would then be devoted to sorting, pinning, and identifying the booty, and conscientiously penciling everything into notebooks, an oasis of bliss before the dreary school week began again on Monday morning.

There are about 4,000 species of beetle in the Netherlands, and I had set myself the task of finding as many of those as possible in Voorne. After two or three years, in the racks of mothballed insect drawers in my room, I had amassed more than 800 different species, some never found before in the country.

The first few hundred of those species were easy: common widespread ones that I would simply bag as they ambled across the path or sat perched on the tip of a leaf. But as my list of catches grew, more advanced collecting techniques were called for to add the more elusive species from so-called "special habitats." Such as myrmecophiles—animals whose place in nature is inside ant nests. My entomology handbook told me that the best time to find these was in the middle of winter, when all the inhabitants of an ant nest would be huddled up in the deeper reaches, and—more importantly—would be too hypothermic to be bothered to bite me.

So, one frosty winter morning, I tied a large spade to the frame of my bicycle and headed for one of the stands of pine in the inner dunes where I knew there were large, dome-shaped nests of the red wood ant, *Formica rufa*. The mounds were still there, covered with the dried-out stems of stinging nettles that had sprouted on top of the ammonium-rich sites. I plunged the spade deep into the ant mound. Heaving up spadefuls of pine needles mixed with ice crystals, I finally reached the frost-free depths where the ants were hiding. I took out my seasoned beetle sieve, a clever time-honored contraption of German design consisting of a strong cloth bag with a sieve and a funnel, and passed hand-fuls of the nest material into it, shaking vigorously to separate the insects from the larger debris, and finally placing the flow-through into a large white plastic sorting tray. Then, I sat down and waited.

Before long, the undercooled ants slowly began unfolding and stretching their legs and unsteadily started walking around on their plastic floor. But they were of no interest to me. What I was after was what I spotted scattered in between the ants. Here, a small brown clown beetle, with its legs held tight against its round glossy body, looking for all the world like a seed. There a ditto rove beetle, its abdomen curled up in alarm. These were the ones I was after! Myrmecophilous beetles, never seen outside of ant nests. I put the beetles in my killing jar (an old jam jar with tissue paper and a few drops of ether), took them home, and care-fully pinned them, adding to the pin a card with a specimen of the ant glued onto it (as recommended in my authoritative beetle book). Then I took out my identification keys to confirm that I had indeed found a whole series of beetle species that I would never have seen had I not taken the trouble to dig up an ant nest in the middle of winter.

In their hefty, definitive volume *The Ants*, esteemed ant-specialists Bert Hölldobler and Edward O. Wilson devote an en-tire chapter to the animals that shack up with ants. They provide a "summary" table that goes on for fourteen pages and covers not just beetles, but also mites, flies, butterfly-caterpillars and spiders. Woodlice, pseudoscorpions, millipedes, springtails, bugs, and

crickets . . . In almost any group of creepy-crawlies, there are species that have crept and crawled their way into the ant society and
found tricks to eke out a living there.

Those tricks are of two kinds. The first is to blend in. Ants live
in a largely chemical world. Communication within an ant society happens with a whole bouquet of scents and smells, with
which ants pass messages to one another that are the pheromonal
equivalents of a simple "Howdy," a comforting, "Fine, fine, everything is hunky-dory," an excited, "Ooo, nice food two leagues west
of nest," or a frantic, "SAVE YOURSELVES!!! SOME BASTARD IS
STICKING A BLOODY SPADE INTO THE NEST!!!"

The ants' chemical language also functions as a social immune
system: it distinguishes "self" from "foreign." Any creature that
does not smell like a fellow colony member is mercilessly attacked.
So, to invade a nest, myrmecophiles (even those that do not mean
the ants any harm) have needed to break the ant's identification
code. They have evolved to speak "ant" to avoid detection. Many
myrmecophiles have special glands on their bodies that produce
their host's signal molecules (especially "appeasement" signals),
which are wafted into the air via tufts of hair. Some myrmecophiles, such as the rove beetle *Lomechusa*, are even bilingual: in
winter, *Lomechusa* lives in a nest of the red stinging ant *Myrmica*
and chemically chats along with them happily. But in spring, it
leaves *Myrmica* and takes up summer residence in a red wood ant
nest and somehow, seamlessly switches its chemical vocabulary
to *Formica*.

The second trick that myrmecophiles have evolved to maintain themselves in ant society is to find a niche where they can be
happy and safe. The ants' obsessive-compulsiveness helps this.
Whenever we accidentally snatch a peep into one when lifting a
rock in the garden, the inside of an ant nest may seem a chaos of
criss-crossing ants and randomly strewn brood. However, it is

actually a highly structured society with dedicated areas for the different services that make the society tick—not unlike a medieval city. There are refuse areas where the colony's waste is dumped; peripheral nest chambers and guard nests where the nest's defensive troops reside; storage chambers for keeping supplies; brood chambers with separate sections for pupae, larvae, and eggs; the queen's private quarters . . .

Some ants have stables where they keep the aphids they milk or vegetable plots for growing edible fungus or for germinating tough seeds so that they can be eaten. And then there are the different parts of the nest's transportation system: foraging trunk routes, thoroughfares within the nest, peripheral branches, even an endlessly branching system of roads connecting the nest with its hinterland; without any central planning or budgets, ants are able to build sophisticated travel networks that human urban planners often cannot match.

Each of these many different substructures of the ant nest and its surroundings has its own specialized myrmecophiles. This already starts on the access roads leading in and out of the nest. The European jet ant (*Lasius fuliginosus*) has its main transportation routes up and down tree trunks, and this is where the beetle *Amphotis marginata* hangs out. These beetles are true highwaymen. By day, they hide in shelters along the trail, but at night they come out and stop passing ants that are returning to the nest with food. The beetle uses its short, powerful antennae to tap the ant's mouth and drum rapidly on top of its head. This mimics, in a rather persuasive way, the begging behavior of ants in the nest, and the startled ant will void its crop, the regurgitated food being quickly lapped up by the beetle. The ant, however, often realizes its mistake and then tries to attack the vagabond. But *Amphotis* is flat and big and heavily armored, and it just cowers, withdraws its appendages and is as impregnable as a tank, so that the duped worker ant soon gives up and returns to its nest empty-handed.

Inside the nest of the jet ant we find another beetle plying its trade. The larvae of the rove beetle *Pella funesta* are the nest's garbage men. They live in the nest's refuse heaps where they consume dead ants, staying out of sight by feeding from below or even getting inside the ant corpses. When a worker ant attacks them, the larvae lift their abdomen which carries glands with chemicals that instantly relaxes or confuses them—like some sort of "antnip." The adults of *Pella funesta* scavenge on dead ants, too, but in addition, they also hunt live ants, sometimes in a group. Like a pride of lions, the beetles give chase and one of them will try to climb on an ant's back, get its jaws into the ant's neck and sever its nerves and throat. These attacks often fail, but if successful, the whole pack of beetles will feast on the prey communally.

The nest's Eldorado, however, is the brood chambers. Here, the ants bring their highest-quality food (freshly-killed insects, for example) for their newborn larvae. Many myrmecophiles have found their dream niche there, either begging food from the ant workers by chemically pretending to be ant larvae, or preying on the larvae themselves. But brood chambers are also heavily defended. Any interloper discovered there will be killed instantly. So the myrmecophiles who have evolved brood chamber specialization have also needed to evolve very sophisticated techniques to evade the ants' enemy detection. The peculiar beetle *Claviger testaceus* is one of them. It bears the hallmarks of millions of years of adaptation to living inside ants' nests. It is pale, with a curious elongated, eyeless head, strange, club-like antennae, and thick tufts of golden hair on its back. Once again, the secret lies in those tufts of hair. Underneath lie glands that produce chemicals that apparently give off the smell of death. Of insect cadavers, that is. An ant worker coming across a *Claviger* beetle will take it for freshly killed prey (further fooled by the beetle playing dead), pick it up by its conveniently stalk-like forebody, and then carry it to the brood chamber, where all the most tasty morsels go. There, it may dump additional bits of decaying meat on the beetle, cover the pile in puked-out saliva with digestive enzymes, and move on to other chores—thinking it has done the developing larvae a favor. But, in fact, as soon as the *Claviger* scrambles from underneath the pile of insect remains, it will start feeding on the ant eggs, larvae, and pupae.

Claviger testaceus, *Pella funesta*, and *Amphotis marginata* are just three of the 10,000 or so different myrmecophile species that scientists think exist, belonging to at least a hundred different families of invertebrate animals. This evolutionary explosion of myrmecophily has probably been going on for as long as there have been ant societies—at least some 75 million years. The reason being that ants belong to that elite corps of movers-and-shakers that ecologists call "ecosystem engineers."

The term "ecosystem engineer" was coined in an article in the journal *Oikos* in 1994 by three ecologists: Clive Jones, John Lawton, and Moshe Shachak. They write: "Ecosystem engineers are organisms that [. . .] modulate the availability of resources to other species, by causing physical state changes in biotic or abiotic materials. In so doing they modify, maintain and create habitat." To put it bluntly: ecosystem engineers create their own ecosystems. It is easy to see how ants fit this definition. Ants branch out into their environment, and, by virtue of their advanced levels of self-organization, concentrate resources in their nest. The inside of the nest is a novel ecosystem with a constant influx of energy in the form of food carried in by ants, that may be exploited by other species. Those 10,000 myrmecophiles are the new species that have evolved to make use of the opportunities that the ants' engineered ecosystem offers. But even species that do not qualify as myrmecophiles may be affected by the ants' modifications of their environment. Such as those stinging nettles growing on the nitrogen-rich patch around the red wood ant nest that I excavated.

Many organisms besides ants are important ecosystem engineers as well. Think of other animals that create structures much larger than themselves, such as termites or corals. And ecosystem engineers need not be tiny. Take beavers, for example. There is no better hydrological engineering team than a family of beavers. They chew down trees and use these, together with rocks, to build dams up to hundreds of yards long. In slow-flowing water, they will build a straight dam, but in a faster-flowing river, the dam will be curved to better withstand the push of the water. The dams cause the stream waters to slow down and widen, creating swamp land that is less easily traversed by the beavers' predators, such as wolves, and maintains a steady supply of beaver food (water plants and tree saplings) during winter. The animals dig canals to transport logs that are too heavy to drag over land, and they build lodges: large hut-like dwellings constructed from branches, twigs and grass, and solidified with mud, bits of wood, and bark. Because of all this environmental enhancement, beavers have such an overriding impact on their environment that they create new niches for whole swathes of other species. Even after beavers abandon an area and the dams they have erected decay and are breached, the resulting flood allows the development of meadows that can persist for decades after the beavers have left.

One area where beavers had such an impact in the past is a large island on the east coast of North America, in the estuary of the Muhheakantuck river. The elongated island has gently rolling elevations and depressions—its local Lanape name means "island of many hills." Until a few hundred years ago, most of those were abundantly clad in chestnut, oak, and hickory forest, which sucked up the abundant rainwater and only released it piecemeal, allowing 62 miles of slow-flowing creeks and streams to develop all over the island. Beavers were aplenty in this fine beaver habitat. In one spot in the southern part of the island, two creeks converged on a gently depressed valley. Beavers dammed the creeks, and the valley transformed into a red maple swamp, slowly colonized by other animals that feel at home in such a habitat, like wood ducks, green frogs, and brown bullhead. Besides the red maple, there was northern water plantain and marsh blue violet. We know all this because of a study—groundbreaking in more than one sense of the word—led by landscape ecologist Eric Sanderson of the Wildlife Conservation Society in New York. Using information on the island's climate, soil types and topography, early Dutch and English records of its landscape and wildlife, and computer modeling of the entire food web of that part of North America, they were able to reconstruct what the landscape, and all the life it supported, looked like four hundred years ago.

Today, nothing remains of what was there. For that island is Manhattan, and Eric Sanderson's work is also known as the Mannahatta Project. The project's purpose was to create a website with a navigable map of today's Manhattan where any location could be chosen, stripped from all its human structures to reveal, in full color and detail, the model's best estimate of that location's habitat and abundant wildlife before Europeans set foot. "[Four hundred] years of development have rendered this earlier abundance as difficult to imagine to us as perhaps our modern roads, skyscrapers, and wealth would be to those first European colonists and their Native American neighbors," Sanderson writes. His aim was reached by September 12, 2009, the quadricentennial of the day when Henry Hudson, sailing in on a ship of the Dutch East Indies Company, first set eyes on it, and scribbled in his log, "[A]s pleasant a land as one can tread upon."

Indeed, when you visit the project's interactive map on http://welikia.org, it's as if Google Earth has directed you to one of the few remaining unspoiled wildernesses on earth. Coast-to-coast forest cover, only interrupted here and there by meadows, swamps, brooks, some Lanape settlements, and a few beaches and rocky bluffs along the shore. A paradisiacal place. But click on the button "STREETS," and the modern street plan appears on top of all that verdure. Suddenly you realize that that lush brook you have been staring at is actually in what today is Harlem, or Greenwich Village. For example, that confluence of two creeks, where beaver ecosystem engineering had created a red maple swamp, lay smack in the middle of what now is Times Square, with one stream appearing from the New York Post building, the other from under the Jacqueline Kennedy Onassis High School.

By now, you may have an inkling of where this narrative has been heading. By clicking on the buttons on the Mannahatta Project's interactive map, we are toggling between the work of one ecosystem engineer and another. The beavers of Mannahatta are gone, but they have been replaced by what we could call nature's ultimate ecosystem engineer: *Homo sapiens*—running around in modern-day Manhattan, the ecosystem it has engineered for itself, like ants in an anthill. And, as with any good ecosystem engineer, in so doing it has created niches for cohabiting animals and plants. Not myrmecophiles, but, if you will, anthropophiles. It is those anthropophiles and the niches they carve for themselves in the human-engineered ecosystem that we will discover in this book.

2
THE ANT(HROPO)-HILL

B Y CALLING *HOMO SAPIENS* NATURE'S
ultimate ecosystem engineer, I used the word "nature"
deliberately, because a crowded, noisy, polluted, concrete metropolis is not what we normally think of when we hear the term
"nature." What we rather think of is something akin to what I
happen to be overlooking as I write these words.

I am sitting on the veranda of a field research center in Malaysian Borneo, where I am spending a few days to prepare for a
tropical biology course. Five yards away from me begins the
undisturbed rainforest. In my field of view there are probably a
hundred different plant species: heavy-duty rainforest trees with
buttressed roots, and with various ferns sprouting from pockets

of debris on their branches, vines and spiny climbing palms, some with nests of *Myrmicaria* ants clinging to them. Over the past two hours, my focus on writing this text has faltered many times and I would stare into that verdant vegetation; I have seen two bearded pigs oinking along, a giant squirrel, a white-crowned shama, at least twenty species of butterfly, a large metallic-green chafer that whizzed past in a straight dash, and I have heard the unmistakeable calls of helmeted hornbills (quickening series of "woohoos" culminating in a maniacal cackle) and a great argus pheasant ("wow-wow!") calling in the distance.

This forest is completely untouched, except for the pegs with colored flags that the students have been sticking in the ground to mark their study plots. In the distance, the forested slope inclines, rising to the 5,000-foot-high, 15-mile-wide circular quasi-crater that was unknown until 1948, when a pilot almost crashed his aircraft into the steep rocky escarpment that forms its rim. It is suspected that this "lost world" never saw any human habitation until this field center was built. If anything is nature, then this is it: wild, unspoiled, free from any form of human corruption.

And yet, when talking about nature, why do we always implicitly or explicitly factor humans out of the equation? Why consider that ant nest hanging in that tree over there as natural, but our human cities not? Why do we admire the leading part that these ants play in the ecological workings of their bit of rainforest, but at the same time do we express disgust at the way humans may dominate a landscape? There is no essential difference. Those ant ecosystem engineers build their nests from materials they obtain from their environment—just like humans. Their societies grow and their workers, who only have the welfare of their nest in mind, harvest anything that is edible from the patch of the earth where they live—just like humans. Given the chance, the colonies will multiply and thrive for as long as their environment can supply them with food and building materials. Just like human cities do. So why do we think of ant society and its role in the global food web as natural, but view human society as an unnatural and unwanted imposition on that same food web?

Rivers of ink have already flown in the many attempts of philosophers, ecologists, and conservationists to define nature and the natural, so I will refrain from releasing my own tributary. But let it be clear that I consider human cities as a fully natural phenomenon, on a par with the mega-structures that other ecosystem engineers build for their societies—the only difference being that whereas ants, termites, corals, and beavers have been maintaining their roles at a stably modest level for millions of years, the scale of human ecosystem engineering has grown by several orders of magnitude over just a few thousand years. Whether, as a species, we are suited for living in such dense, complex communities is another matter, and I will come back to that at the end of the book. But first, let us examine the modern, human megacity for what it is: an exciting, novel ecological phenomenon.

In the beginning, at a time when our species had just emerged from its smaller-brained predecessors, and was still so rare that it would have qualified as vulnerable by today's Red List standards, we were already small-time ecosystem engineers. Not unlike beavers, our hunter-gatherer ancestors would find a suitable spot, preferably with natural shelter—a rocky overhang or a cave, perhaps—and settle there for a while to exploit the environment before moving on. Some "proto-domestic" animals like the ancestors of dogs may have followed in our wake and hung around camp to scavenge refuse, and we may even have brought our own domesticated animals and plants: edible rodents in cages (like the Polynesian rats carried around by the Lapita people) or cuttings of medicinal plants. Settling would involve burning or clearing the vegetation around the camp, tending edible and medicinal plants and weeding away the undesired ones. We would create fireplaces to cook the fish and game we would hunt, as well as the clams and snails we would collect from the stream. We would raid bees' nests to get honeycombs and the protein-rich bee brood, hunt the local megafauna, and collect fruits and nuts from the forest. Like beavers, we may even dam a creek, in our case to collect the fishes that would then be splattering about in the shallow waters downstream. Our effects on the environment would be subtle—a drier microclimate due to clearing of the vegetation, a local depletion of large animals, the introduction of a few alien species—and the environment would quickly recover whenever a group would pack up and leave for new hunting grounds.

Much of this changed when we took up farming. The revolutionary invention of growing food, rather than searching for it, had two important consequences for our way of living. First, growing crops for subsistence around a settlement meant that it was no longer necessary or profitable to live the nomad life. After all, going to the trouble of setting up fields and planting them clearly was a long-term investment. Until the soil became exhausted, it was best to stay put. Secondly, it meant that our trophic level changed—trophic level being an organism's position in the food pyramid. Green plants that use the sun's energy and "eat" carbon from the air sit in trophic level 1 as the world's chief "primary producers." Level 2 is occupied by vegetarian animals that consume those primary producers. The third level of the food pyramid is where we find the predators that eat those plant-eaters, and so on. The food pyramid is pyramid-shaped because only one-tenth or so of the energy produced by one layer is carried upward into the next layer. The rest is lost along the way as waste, heat, and power to run the bodies of the organisms in that next layer. And since energy translates into how much life a level can support, in any habitat you'll find ton of green stuff (level 1), millions of plant-eating insects (level 2), thousands of insectivorous birds (level 3), a bunch of weasels and hawks (level 4), and perhaps just a single top-predator, like a lone tiger or a solitary eagle, at level 5. Humans, by changing from chiefly hunting to mostly farming, collectively stepped down one tier in the trophic pyramid—where there was much more energy and therefore much more space to grow.

And grow we did. Five or six thousand years ago, we improved irrigation and soil tilling to such a level that frequent relocation because of depleted soil nutrients was no longer necessary. Agriculture proved such a success that not everybody in a village needed to be involved with it. It was left to specialists while the rest of the settlement could take up other necessary trades. This meant that these permanent settlements became places that could supply food and coveted goods to their hinterland. This in turn led to the development of transportation technologies, and people skilled at building and maintaining these. Cities also became the places from which organized warfare emanated, subjugating tribes that still adhered to hunting and gathering, and thus further spreading the agricultural, village-building lifestyle. Around that time, roughly 6,000 years ago in Mesopotamia, the first true cities appeared. One by one at first, but as the centuries went by, more and more parts of the world began to show signs of urbanization, with new cities popping up in India and Egypt, then in more rapid succession: Pakistan, Greece, China . . . An animation based on the research by Meredith Reba and her colleagues at Yale University shows how cities appear all over the planet, from 5,700 years ago until today—slowly at first, but then,

like corn popping in a pan, building up to the deafening cre-
scendo of urbanization during the past century.

Over the next few decades, the popping is only expected to
become louder, with megacities (10 million inhabitants or more)
setting the stage. In the Pearl River Delta, one of China's main
economic hubs, so many cities are now squeezed together in an
area smaller than the unimpressive size of Belgium, that it is
termed a "megalopolis" with a combined population of 120 million, almost the same as all of Russia. By 2030, almost 10 percent
of all people on earth will live in only 41 megacities, and most of
those will be in Eastern China, India, and West Africa. Kinshasa,
still a quiet backwater a few decades ago, will hold 20 million
people, and in Lagos the population will be more than 24 million.
Those figures may be mind-boggling, yet the strongest urbanization, relatively speaking, will actually take place in the small and
medium-sized cities (that's anything below 5 million inhabitants) in formerly rural countries. Such cities are expanding rapidly, by over 2 percent a year, while the annual growth rates of
the really big megacities is just 0.5 percent. Over the next decade, the developing world's smaller cities are going to absorb
twice as many people as their bigger brothers. Between 2000 and
2010, the urban population of a country like Laos, for example,
which lacks really large urban centers, has doubled.

All these statistics are not to say that experts agree over what a
city really is. The socio-economic definitions vary from period to
period and from place to place. In Norway, a settlement with 200
inhabitants is already considered urban, while in Japan you need
50,000 for the same status. City status may also be an administrative thing. Some cities are "official" and therefore can claim certain benefits from the state. For example, only two of the twelve
London boroughs are official cities, while none of the others, nor
London as a whole, is legally entitled to call itself a city. Not to

complicate matters, I'll take a pragmatic approach and simply consider cities as those areas where the density of humans and their buildings are distinctly increased and with it, the infrastructure and average income. But those are only the human factors. In their wake come interesting ecological characteristics.

3
DOWNTOWN ECOLOGY

"PANG!" SOW-YAN USES BOTH HANDS, one pulling an invisible trigger, the other holding an imaginary barrel aimed at the blazing mid-day Singaporean sky, to imitate a rifle shot. Again he goes, *"pang!!."* The impersonation comes in reply to my question about the Indian house crow situation. "In my area, they shoot them down," he elaborates with some indignation. "For no reason! Just someone complains about them, and that's it. Also, everybody is using the wheelie-bin now, so the crows cannot get to the garbage anymore. Last time, they would just tear open all the refuse bags."

We are hiking along Singapore's south coast. My host, Chan

Sow-Yan, a retired computer engineer, naturalist, and local mollusc expert, has paused for a second to do his crow-culling dem-
onstration and then paces on, toward the place where the Rochor
Canal joins the Kallang River. Here, a promontory juts into the
waters, and he takes me up there to overlook the estuary. The
group of house crows (*Corvus spendens*) has flown off, but their
place is immediately taken by an excited bunch of Javan mynahs
(*Acridotheres javanicus*), beautiful anthracite-and-white birds with
mischievous eyes, bright yellow legs and a ditto beak adorned with
a small crest of tufty black feathers. The mynahs begin running
about, picking up edible morsels from among the cow grass
(*Axonopus compressus*) and touch-me-not (*Mimosa pudica*). Sow-Yan
points at the water's edge, where the cow grass gives way to yellow-
flowered water mimosa (*Neptunia oleracea*). Then, pointing left
and right, he draws my attention to the clumps of pink eggs of the
Pomacea apple snail clinging to the shore, the massive peacock
bass (*Cichla orinocensis*) coming up for air, and a tiny red-eared
slider turtle (*Trachemys scripta elegans*) quietly paddling along just
beneath the water surface.

Kallang Riverside Park is a rich tropical ecosystem. But that
does not mean it is a wild, idyllic paradise. Instead, it's a tiny
pocket of greenery tucked in between Singapore's high-rise buildings. A few lawns with clumps of mango, coconut, and fig trees;
Malay girls taking selfies on benches, and winding paths where
European joggers brush shoulders with Indian youths on skate-
boards; a helmeted elderly Chinese lady on a bicycle with three
coconuts in the front basket. The promontory where Sow-Yan and
I stand, and the embankments with the patches of pink snail cav-
iar, are made of unforgiving concrete. The river is no longer tidal
because of the gigantic Marina Barrage downstream. The mynahs
and crows are eating from discarded coconut shells and other
leftovers from people's picnics, and the carpet of freshwater algae

that the turtles and water snails forage on grows over bricks and plastic bottles. Due to flooding or leaks of the city's sewage system, the water itself carries the unmistakeable chemical signature of the 5.7 million inhabitants of Singapore. A study led by Xu Yonglan of Singapore's Nanyang Technical University found 0.1 milligram of pharmaceuticals per liter of Kallang River water (mostly painkillers like ibuprofen and naproxen); similar amounts of estrogens (from cosmetics and pharmaceuticals) and an insecticide that is used to kill fleas and ticks on pets. In other parts of Singapore, the researchers found up to 1.2 milligrams of caffeine (about as much as in a teaspoonful of coffee) in every liter of river water.

Moreover, each and every one of the animals and plants that Sow-Yan and I spot are not native to Singapore. The house crow, originally from India, Sri Lanka, Myanmar, and Yunnan, suddenly appeared in the port in 1948. Nobody is sure where they came from. Perhaps plantation-hopping from Malaysia where they had been released half a century earlier to control a caterpillar infestation in coffee groves. Or maybe as stowaways on ships. Either way, the crows fared well, increasing from several hundreds in the 1960s, to hundreds of thousands by the beginning of the twenty-first century. Despite the culling of at least 300,000 crows over the past fifteen years, and a slew of measures to discourage them from foraging in refuse and nesting in the omnipresent yellow flame trees along Singapore's streets, the crows remain a common sight (and, according to Sow-Yan's neighbors, a nuisance) all over the city. The Javan mynahs arrived as pets (they are sought-after cage-birds, and virtuoso impersonators) around 1925, from Java or Bali, where they occur naturally. In the 1960s, ornithologist Peter Ward still wrote of them, "A shy bird which visits gardens in the suburbs, but which is only occasionally seen in the city." Since then, they clearly have thrown all shyness to the wind and have become the city's most numerous and noisiest bird species, probably rivaling the human population in terms of numbers. "The chairs in the coffee shops are full of their shit," Sow-Yan says ruefully.

The ubiquitous cow grass, a tough broad-leaved grass that is underneath almost every footstep you take in Southeast Asia, is originally from Central and South America, and the same applies to its companion, the ever-entertaining touch-me-not, with its leaves which wrap up instantly when you touch them. Their sticky seeds have been hitch-hiking their way across the globe for centuries via clothes, shoe soles, and the wheels of vehicles. And nobody is really sure where the water mimosa is from, but it's certainly not native—probably an introduction from Mexico, Sow-Yan thinks.

The huge apple snails that we see slowly sliding along the sheets of plastic at the bottom of the canal, their tentacles bristling, originally hail from South America. They began their global maraud probably via dumped aquarium water and are now proud members of the snail contingent on the list of the world's most feared invasive alien species. Also on that list, and another escapee from the aquarium trade, is the red-eared slider, again a species originally from tropical America, while the peacock bass, who can call the Amazon river its home, has settled in the city state thanks to "overenthusiastic angling fans," according to Singaporean fish experts Ng Heok Hee and Tan Heok Hui.

The urban ecosystem of Singapore, like that of cities all over the world, no longer consists of a selection of local, native species. Instead, in tune with its human population, it has been assembled from immigrants from all over the globe. Either intentionally or accidentally, people have been ferrying flora and fauna across the world for as long as they have been trading and traveling. Places where human activity reaches fever pitch—places like Singapore, the world's second-largest port—abound with such exotic species. These urban ecosystems are formed not by ages of evolution or the slow colonization by species under their own steam and of their own choice, but by human diligence alone. Many cities' ecosystems run completely on non-native species, especially under water. The marine environment in San Francisco Bay, for example, is dominated by sea creatures from elsewhere. Most of these have probably hitched rides in ballast water—the seawater (including everything that lives in it) that ships pump into their hulls to improve balance after unloading their cargo, and which they then dump in their next port-of-call.

Sow-Yan observes the beads of sweat that have begun to form on my brow. "Thirsty? Shall we have a drink?" He leads me across Crawford Street into a maze of tall apartment buildings. We stop in a small square, planted with cow grass and a few ornamental palm trees inside a towering chimney composed of multiple blocks of flats, their hundreds of growling air-conditioning units billowing out a constant stream of hot air. We sit down at an open-air food court and watch the Javan mynahs forage for scraps of food among the legs of the plastic tables, their beaks slightly ajar in an attempt to cool off. They, too, are feeling the effect of the urban heat island.

First described by geographer Tony Chandler in his 1965 book *The Climate of London*, the "urban heat island" is the result of several things. To begin with, the activities of millions of people, packed together in a small area with all their cars, trains, and other machines, creates a lot of excess heat, which remains trapped among the tall buildings. Secondly, the stone, asphalt, and metal of streets, pavements, and buildings absorb heat during the day, either directly from the sun or via reflections off windows, and at night only slowly cool off, radiating out heat all the time. The bigger the city, the larger the heat island; every tenfold increase in number of inhabitants raises the temperature by about three degrees centigrade. In the world's largest cities, it can be more than twelve degrees hotter than in the surrounding countryside. Furthermore the column of hot air that slowly rises from the city center provokes a city-directed breeze from all directions. As it rises, the air column cools down and water begins to condense around particles of city dust contained within it, resulting in a phenomenon called urban rain. In other words, some cities are so large that they produce their own climate: the wind is always blowing toward them, and it's distinctly hotter and wetter there than in the surrounding rural areas.

In Singapore, the urban heat island has its epicenter exactly where Sow-Yan and I are sweating over our sugar cane juice. According to measurements by the National University of Singapore, we may add seven degrees Celsius to the already balmy tropical temperature. It's time to get into Sow-Yan's air-conditioned Toyota and head for the Marina Barrage.

To get there, we have to travel some four miles around the Downtown Core, the futuristic heart of Singapore, where massive feats of modern architecture are bathed by ten-lane expressways, like boulders in a stream. From the car, I glimpse forgotten bits of vegetation between the buildings that remind me that urban ecology is an ecology of fragmentation. Much of the city consists of concrete or steel, surfaces that can only support rock-perching birds like swifts and peregrine falcons, and minuscule life forms that create a film-like ecosystem on the weathering surface (bacteria, lichens, algae—and some tiny animals such as silverfish and springtails that can eke out a living in this two-dimensional habitat). Most other life forms cannot live on the city's impervious surfaces, but need some sort of soil. Mind you, "soil" may be as simple as those cracks in the pavement where the airborne spores of a *Pteris multifida* fern germinate. Or the edge of a drain where the seeds from a discarded starfruit take root and begin retaining moisture to lay the foundations for a miniature ecosystem suited to nematode worms, ants, and mosses. It can be a few square yards of actual vegetation—the raintrees along the roadside, the potted plants on balconies, the thickets of vines and creepers climbing the legs of the Ophir Road flyover . . . even the roof garden of the newly built megalomanic Marina Bay Sands Resort that looms in the haze like a latter-day Stonehenge. Or, the few larger greenspaces: pocket parks like the Kallang Riverside Park or relics of rainforest, such as the Bukit Timah and Central Catchment Nature Reserves. A glance at Singapore's map shows mostly forest fragmentation: scattered flecks and wisps of green among large swathes of the gray and brown of built-up areas.

Of the original 208 square miles of rainforest that clad the island two centuries ago, when the Sultan of Johor allowed the British to pitch their imperial tents, less than one square mile remains (in Bukit Timah and the Central Catchment). In addition, there are some eight square miles of "secondary" vegetation: the stuff that makes up most of those postage-stamp-sized bits of green on the map. Any organism that cannot live its entire life on bare concrete will need these green islands and islets to maintain itself in the city.

But there's a thing with islands—the smaller and more isolated they are, the less life they support. In the 1960s, entomologist Edward O. Wilson and theoretical ecologist Robert MacArthur famously crafted a new ecological theory that they called "island biogeography." It went like this. Imagine a bunch of islands. These could be real islands in the sea but also any other fragments of habitat. The number of species (of, say, butterflies) that live on each island depends on two things: how many different butterfly species manage to reach the island, and how quickly butterfly species tend to become extinct there. The smaller the island and the farther away it is from the mainland, the more likely it is for a butterfly to flutter by and not settle there. But when a species does colonize it, its survival also depends on the size of the island. On a big island, the population can grow to, perhaps, thousands of individuals and the survival of the species is quite secure. But on a small island there may only be space for two dozen individuals of the species, and a heatwave or a disease could easily wipe them out. All these effects combined, Wilson and MacArthur discovered, produce a set of mathematical rules that make the number of species on an island surprisingly predictable. Roughly, with every tenfold increase in island size, the number of species you find there doubles. This goes for butterflies just as well as for beetles, bugs, and birds.

A large town or city, with its archipelagos of green in oceans of asphalt, is an island biogeographer's paradise. In the town of Bracknell in the UK, for example, ecologists studied the Hemiptera (a group of mostly plant-feeding insects including "true" bugs, aphids, and cicadas) living in the circular bits of vegetation in the centers of traffic roundabouts. These road islands lying in seas of tarmac followed island biogeographic theory to the letter, showing a perfect relationship between roundabout size (which ranged from 4,300 to nearly 65,000 square feet) and numbers of hemipteran species.

Constructing road islands is one way in which cities can create archipelagos. But expanding cities also create islands by shredding up existing forests. This is one reason why urban ecosystems contain just a small subset of the species that lived there before. In an article in *Nature* in 2003, Australian ecologist Barry Brook, with Navjot Sodhi and Peter Ng of Singapore's Lee Kong Chian Natural History Museum, enumerated precisely what had happened to Singapore's flora and fauna since the start of urbanization in the early nineteenth century. Thanks to Victorian-era collectors such as Alfred Russel Wallace and Stamford Raffles, as well as learned societies like Singapore's Nature Society (founded in 1954), we know a lot about the city's natural history. Indeed, much of its nature *is* history. As Brook, Sodhi, and Ng found out, over the past two centuries or so, as the erstwhile continuous rainforest was logged, converted, and fragmented, species disappeared from the island one by one. The huge tiger orchid, the largest orchid in the world, was last seen around 1900, while tigers themselves left the island for good when the last one was shot in 1930. The Great Slaty Woodpecker disappeared in the second half of the twentieth century. Today, depending on the type of plant or animal, 35 to 90 percent of the original species has gone, or survive only under strict husbandry in Singapore Zoo and the Singapore Botanic Gardens.

We park the car at the Sustainable Singapore Gallery and walk across the Marina Barrage to the Marina East Park on the other side. There, we take the concrete track that winds among the recently cut turf. Large dragonflies are zigzagging above the grass, targeting the clouds of midges that are beginning to gather in the setting sun. A public greens worker in an orange vest and a paddy hat uses his smartphone to document his impeccably mowed lawn, then gets on his all-terrain bike and cycles off. Here and there, the dried and broken carcasses of large black-and-yellow millipedes lie on the cycle path, the victims of failed attempts to cross the scorching concrete. It's *Anoplodesmus saussurii*, another exotic species, says Sow-Yan.

We take a right on a sand path that leads through a strip of coastal scrub onto an expanse of reclaimed land, offering a view of the scores of ships moored offshore. A group of bird-watchers with telescopes and binoculars is gathered at the far end of the spit of land. "We have about 2,000 bird-watchers in Singapore,"

Sow-Yan says. "Also a few hundred people working on butterflies
and dragonflies. And some shell enthusiasts, but not enough."
Sow-Yan pulls out his binoculars. "What are they looking at?" he
mumbles, as he watches the bird-watchers closely. "Ha! They're
watching a house crow!"

Twelve urban twitchers with high-end equipment observing
a single invasive urban bird. It's a familiar sight among urban
naturalists all over the world. Like anybody else, biologists,
whether professional or amateur, tend to live in cities. It is there
that we also find the libraries, natural history collections, and
nature clubs. With such concentrations of knowledge and interest in biodiversity, it is perhaps not surprising that the city is one
of the best studied habitats in the world. It is also where emo-
tions about our fellow species tend to flare. As the house crow
guides us into the world of urban nature study in the next chapter, prepare for a tale of passion, tragic death, and politically mo-
tivated killings.

4

URBAN NATURALISTS

S INGAPORE IS NOT THE ONLY CITY IN
the world to have been invaded by house crows. Humans
have been transporting them throughout much of the tropics,
either intentionally (as honorary "garbage-collectors" or for pest
control) or accidentally, as stowaways on ships. Besides Singapore, they now live in many other countries in Southeast Asia, as
well as the Middle East and East Africa. In fact, they don't have
a non-urban habitat anymore, and are exclusively found in towns
and cities in the tropics. As bio-philosopher Thom van Dooren
writes, "You might say that, in so far as these birds have a 'natural environment,' we're it."

But in 1994, something momentous happened. A male and

female house crow turned up 52 degrees north of the equator, in the port of Rotterdam, possibly by way of a cargo ship from Egypt. Surprisingly, the tropical crows survived the cold winter of 1996–97, when temperatures in the Netherlands dropped as far as minus 20 degrees Celsius, and in the following year even produced a nest with chicks. From then on, the population grew into a breeding colony in trees around a soccer field, where the birds built nests lined with colorful nylon strings that they would pull from discarded shipping rope in the port, and fed their chicks with fish and chip scraps pilfered from the portside "Het Vispaleis" fish stall. By 2013, there were about thirty of them, and bird-watchers would regularly come down to the port to add the sleek crow species to their "life list."

It remains a mystery how birds that normally breed in the hottest parts of the globe can suddenly shift their niche poleward. The urban heat island probably helped, and the milder climate near the seashore. But still—we will come back to this and similar mysteries later. Let us first examine the sad fate of this interesting population. Not everybody was as kindly disposed to them as the local naturalists.

The provincial government, for one, certainly was not. Alarmed by the crow's reputation of becoming a pest wherever it settled, it ordered for the birds to be exterminated, much to the dismay of many of Rotterdam's bird lovers. Initially, the plan was successfully challenged by an animal welfare NGO on the grounds that the bird had, somehow, managed to get itself listed as legally protected. The authorities then proceeded to have its protected status annulled and in 2014 a new court ruling cleared the road for a professional hunter to be hired for culling all the birds.

But extermination proved easier said than done. The hunter found the inhabitants of Hoek van Holland on his path, who organized themselves in a "Save the House Crow" committee and vehemently protested against his avicidal intentions. The fact that the hunter accidentally shot some native *Corvus monedula* jackdaws ("they look really, really similar," he complained to the Dutch newspaper *Algemeen Dagblad*) certainly did not help. Moreover, the house crows were smarter than anticipated. As he bagged his first birds, the other crows immediately became wary. "As soon as they see my car, I can already hear their alarm calls. Those birds are so damn clever."

Two years down the line a stalemate seems to have been reached—the hunter trying to outsmart his prey by arriving in his wife's miniature car or by donning a funny red hat to avoid recognition, and a diminishing number of crows managing to cheat death by quick evasive action. Still, after repeated hunts and many an air rifle shot, most of the birds have perished and are now stored in the Rotterdam Natural History Museum. But rumor has it that a few are still at large. However, it is hard to get any reliable information on how many—if any—are left and exactly where they are. The Dutch website waarneming.nl, where naturalists can record their sightings of wildlife, is no longer releasing any for the house crow, so as not to play into the hands of the hunter, and the Facebook group devoted to the crow situation is similarly taciturn. So, when I approach Sabine Rietkerk of the Save the House Crow committee, in preparation for my own jaunt down to Het Vispaleis, I am met with the usual suspicion when inquiring about the whereabouts of a fugitive. In a lengthy exchange on Facebook, I am able to convince her of my good intentions and that I am not a front for the hunters. Initially she will not admit that there are still house crows living in Hoek van Holland, but eventually confides that the few surviving birds are no longer hanging at Het Vispaleis, but have cunningly shifted to a safer place. "They are hiding among the people . . . where the

hunters cannot get at them. Try your luck in the shopping center," she tells me.

So one summer morning I find myself on expedition in the
commercial hub of Hoek van Holland, the portside suburb of Rotterdam and the house crows' 'hood. A few cafés, a newsagent's,
two supermarkets engaged in cutthroat competition, and a liquor
store encircle a windswept square lined with closely-cropped elm
trees. With my binoculars at the ready, I initially spot only jackdaws and herring gulls. But then, as I take a second lap around
the square, one lonely, unmistakeable Indian house crow (identical to the ones I saw in Singapore only weeks before) is stepping
across the road among the shopping-bag-toting pedestrians, right
in front of me. Striding, more like. Long legs with a deliberate setting down of the feet, a graceful metallic-black body and a silvery
brownish-gray hood, high forehead and a long beak. I manage to
sneak a quick snapshot of it before it hops onto the pavement,
takes a rustling bound into the branches of an elm tree, and disappears from sight. Its tree is right next to the terrace tables of one
of the cafés, so I sit down and order a cup of coffee. The crow,
hiding among the foliage next to me, is calling alternatingly in a
raw and in a melodious metallic voice. Via Facebook, I send my
snapshot to Sabine Rietkerk, who replies instantly: "Ah, good
you found it. Yes, that one is often in that spot. He's very vocal.
And isn't he beautiful?"

A few hours later, in the collection depot of the Rotterdam
Natural History Museum, I am looking down into a cardboard
specimen box with the twenty-six stuffed, stretched-out and
neatly labeled Rotterdam house crows that met with the government marksman's lead pellets. Probably siblings, parents, uncles,
and aunts of the one I saw sauntering across the street that same
morning. Lying there stiff and side by side in their glossy black
plumage, they look like body bags after a gang war. "They *are*

beautiful," agrees museum director Kees Moeliker. "It's a sad tale, of course, what's happening there in Hoek van Holland. They're hunted down for political, not ecological reasons. But we're happy that we managed to persuade the authorities to deposit the killed birds in our museum—otherwise they would just have destroyed them, I guess. It's the only European population of this species, after all. Quite special, and splendid material for research."

The crows are the latest addition to the museum's growing collection of urban natural history specimens. Moeliker takes me to a steel rack full of stuffed foxes, wrapped in clear plastic to keep out the insects. Over the past ten years, foxes have begun entering the city from the surrounding countryside, and whenever one is hit by a car, it ends up as a fine specimen in the museum's depot. Of one recent acquisition, the stomach contents were also preserved by the diligent museum curators. They display, in a single five-course meal, this particular fox's transition from a rural to an urban diet: rose hips, a small rabbit, an apple, doner kebab, and cherries in thick syrup.

Conversely, the museum also pays particular attention to species that are disappearing from the city, like the red squirrels that used to live in the Kralingse Bos, the city's largest park, but became extinct in the 1990s. Moeliker holds up a stuffed squirrel, pinned clumsily to a piece of tree branch stuck to a plank. "A few years ago, an old lady brought us this. Normally, we don't really welcome this kind of decorative item, but she told us it had been found dead in the Kralingse Bos in 1966. It's the only specimen we have of the time that the population was still alive and well. So that's kind of nice."

Up in the public exhibition of the museum, the theme of urban nature is unleashed with even more abandon. One showcase holds nests that urban swans and pigeons built from plastic bottles, chunks of polystyrene, chicken wire, and rubber bands—which in some parts of the city are much easier to come by than genuine branches and twigs. Another one has displays of the surprising diversity of moth species found in the city center. Also on view are herbarium specimens of wildflowers that normally grow on the saline seashore, but that now live along the city's roadside verges thanks to winter salting; and of plants whose natural home are rocky ledges in the southern-European mountains, but that now also thrive among the stony walls of Rotterdam's heat island.

The blockbuster, however, is the exhibit *Dead Animal Tales*, a row of showcases in the museum's central hall, which hold meticulously mounted specimens of city animals that collided head-on with their human cohabitants in a particularly memorable way. The McFlurry-hedgehog, for example, is a hedgehog (*Erinaceus europaeus*) that met its demise when its head got stuck in the hole in the top of a plastic McFlurry soft serve ice-cream cup—and is mounted in precisely that undignified position. It is just one of many similar hedgehog victims of this popular fast-food dessert. As the accompanying card explains, "Searching for leftover ice cream, the hedgehogs would get their heads in through the wide opening in the lids, but their spines stopped them from pulling out again. They would die from starvation or blindly walk into water and drown." Another classic is a stuffed house sparrow (*Passer domesticus*) next to a plastic butter tub with the words "domino sparrow" written across in black marker pen. In 2005, this particular sparrow managed to sneak into a hall where 4 million domino tiles had been set up for a live televised event called Domino Day. By the time the panic-stricken sparrow had knocked over 23,000 of them, it was decided that this had to end, and a guy with a gun (in fact, the same professional sharpshooter who is now the house crows' nemesis) did the honors. Once again, the museum's display card text is unsurpassable, so I quote: "The death of the sparrow sparked major commotion (and commotion about the commotion). [. . .] After intense lobbying [. . .] the museum was able to acquire the dead sparrow and the butter tub in which it was kept."

Not only is the museum a central repository for the urban flora and fauna of Rotterdam, it is also a hub for every Rotterdammer who wishes to devote him or herself to one particular sliver of the city's biodiversity. Like everywhere in the world, this demographic is rapidly growing. Cities are full of passionate people who build insect collections and herbariums, or document butterflies, plants, and birds with the cameras on their smartphones, logging their observations on global "citizen science" Internet platforms like Observado or iNaturalist. Or they may be activists and fight for the preservation of urban biodiversity hotspots, iconic ancient trees or rare species. In Rotterdam, there are various nature clubs in town (including single-issue ones such as Save the House Crow). And, says Moeliker, the museum's Rotterdam Urban Ecology Unit maintains a large network of enthusiastic amateur naturalists.

Some of those Rotterdam enthusiasts began as members of the local branch of the Royal Dutch Society for Natural History (KNNV), which was erected way back in 1917. Similarly, many big-city based nature societies all over the world date back to the early twentieth century or earlier. Dates of foundation for the Paris, Belfast, Bombay, and London natural history societies are 1790, 1821, 1883, and 1913, respectively—the urban naturalist is by no means a recent phenomenon. But, as Moeliker's predecessor, former museum director Jelle Reumer, points out in his book *Wildlife in Rotterdam*, an interesting shift took place in nature clubs

worldwide around the middle of the twentieth century. To illus-
trate this, Reumer takes the bibliography of *Mannahatta*, the book
that accompanies Eric Sanderson's Manhattan project that we
came across in an earlier chapter. It lists field guides to New York's
biodiversity from the early nineteenth century until today. Prior
to the mid-twentieth century, Reumer notices, the word "vicinity" almost without fail appears in the books' titles: *Synoptical
View of the Lichens Growing in the Vicinity of New York* (1823), *The
Frogs and Toads in the Vicinity of New York City* (1898), *Plants of
the Vicinity of New York* (1935). But from the late 1950s onward,
such "environs"-indications were dropped from newly appearing titles: *A Natural History of New York City* (1959), *Wild New
York: A guide to the wildlife, wild places and natural phenomena of
New York City* (1997), *Damselflies and Dragonflies of Central Park*
(2001) . . .

It's a clear sign that something has changed over recent decades. Rather than using the city as a comfortable base camp from
which to explore the wild hinterland beyond the city limits, the
city itself has become urbanite naturalists' chief interest. And not
just amateur naturalists. Already in the 1960s and 1970s, a thriving urban biodiversity research group developed around the German botanist Herbert Sukopp at the University of Berlin. In those
cold war days, West-Berlin was an urban enclave of the West in
the largely inaccessible communist eastern Germany, so West-
Berlin's ecologists had little option but to focus on their own city
environment, and they did this with great dedication, leading to
Sukopp's department becoming the cradle of serious urban wildlife studies.

Other countries followed suit. In Melbourne you'll find the
Australian Research Center for Urban Ecology and Seattle houses
the Urban Ecology Research Lab, led by Marina Alberti, whom
we will meet toward the end of this book. Warsaw hosts Marta

Szulkin's *Wild Urban Evolution and Ecology Lab*. The first urban ecology textbooks in English were published in the 1970s in the UK and the US, and scientific journals like *Urban Naturalist* and *Urban Ecosystems* have been around for over twenty years already. There are also global societies like the *Society for Urban Ecology*, which hold annual conferences where urban ecologists from all over the world meet.

So, professional biologists are focusing their attentions more and more on urban habitats; citizen science websites for urban naturalists are popping up everywhere on the Internet; in each large city in the world, books and leaflets are produced that help people get to know their local birds, plants, or insects; and more and more people take high-quality photographs of their local wildlife and use crowd sourcing to have them identified. There are even feature films in cinemas about urban nature, such as the 2015 film *Amsterdam Wildlife* that was screened in six cinemas in the Netherlands.

Through all this activity, we are beginning to learn more about the biodiversity of cities. Sometimes this is thanks to the monastic dedication of individual naturalists, such as entomologist Denis Owen who, for several years in the 1970s, tirelessly maintained a so-called Malaise trap in his garden in the city of Leicester, UK. A Malaise trap is a sort of nylon gauze tent that insects fly into but cannot get out of. They clamber around inside the fabric until they miserably slip into a bottle of alcohol at the top. With his trap, Owen got almost 17,000 hover flies belonging to a total of eighty-one species (roughly one-quarter of all the hover fly species known to occur in the UK). The ichneumonid parasitic wasps that ended up in his trap amounted to an astonishing total of 529 species. For good measure, he also hand-netted and identified all 10,828 (!) butterflies that he saw in his garden. (Butterflies usually don't get into Malaise traps.) Altogether, these belonged to twenty-one species. Owen released every butterfly he caught, and to make sure he did not count the same individual twice, he even took the trouble of giving each one an individual mark written with pen on one of the wings.

Since such near-superhuman dedication is rare, other urban biodiversity "expeditions" are a communal effort. In the 1970s, the Rotterdam KNNV did an inventory of all the insects and plants in a triangular piece of wasteland between three railway tracks in the city center. And since 1996, when the term was first coined for an event in Washington, D.C., the "BioBlitz" is now a household name in urban ecology: a quick twenty-four-hour survey of the biodiversity in a park or some other small habitat, by a large group of professional and amateur scientists. In the US, there is even an annual "City Nature Challenge," where citizen scientists in large cities nationwide (sixteen cities in 2017) try to beat each other in a one-week biodiversity race. Other initiatives are even more playful: the French group *Belles de Bitume* ("Tarmac Beauties") is organizing nationwide "ecological street art": amateur botanists identify wild plants growing on streets and pavements in cities and then scribble down the species' name in colorful, ornate chalked lettering next to them.

Even those who wish to discover completely new species can do so literally by being armchair naturalists. Among the parasitic wasps that Denis Owen got in the Malaise trap in his English garden were two species new to science. When in the middle of October 1995, Mitsuhisa Fukuda bored a pipe into the floor of his house in the city of Uwajima in southern Japan, he pumped up two previously unknown, blind subterranean water beetles from the waterlogged soils under the city. A 2007 BioBlitz in Wellington, New Zealand, turned up a new species of diatom alga. In 2014, two Brazilian mollusc experts discovered a new snail species hiding in the Burle Marx park, a tiny park smack in the city center of São Paolo (one of the largest cities in the world). And in that same year, a new frog species, the Atlantic coast leopard frog (*Rana kauffeldi*) was discovered in the New York–New Jersey metropolitan area, a stone's throw away from the Statue of Liberty.

But could this apparently rich urban biodiversity be just an illusion, caused by the fact that most biologists and naturalists live in cities and simply record more wildlife in the streets where they live and work than elsewhere? The fact that 529 species of ichneumonid wasps are known from Leicester (and not from the neighboring countryside) is, after all, entirely due to Denis Owen's street address. In Amsterdam there is the Amsterdamse Bos, a park that was, in the mid-twentieth century, the favorite playground of beetle expert A.C. Nonnekens, leading to around a thousand beetle species (some 25 percent of the Dutch beetles) being listed for that one city park. Similarly, the city of Brussels boasts about half of the entire Belgian flora—no doubt also thanks to the activities of the large band of Brussels-based botanical Belgians.

Still, this is only part of the answer. For even when ecologists run so-called standardized rural-to-urban transects, sampling random quadrats of land along the gradient from countryside to city center, they usually find that the urban biodiversity dip is not as deep as expected. Indeed, especially for plants and occasionally also insects, it's sometimes a peak.

So what is the biodiversity like, that all this natural history activity is revealing? What kinds of communities of plants, animals, fungi and bacteria are we sharing our cities with? A lot of exotic species, apparently. But also native ones that happen to find in cities something that resembles their native habitat. And species that simply are hanging on in remnants of wild vegetation tucked away in forgotten corners of the urban jungle. But what exactly determines whether a species thrives or perishes in the city? In the next two chapters, we will look into what makes—or breaks—an urban species.

5
CITY SLICKERS

I**T'S 463 PAGES OF DANISH GOTHIC** script. Furthermore, it's a mediocre scan on Google Books that I'm trying to access through the shoddy wifi in the intercity train from Leiden to Groningen. Those are my main excuses for failing to find the first-ever reference to urban plants. According to Herbert Sukopp, the Berlin-based patriarch of European urban ecology, buried in Joakim Schouw's 1823 tome *Grundtraek til en almindelig Plantegeographie* (Foundations to a General Geography of Plants) should be the very first mention of urban plants. I will just take Sukopp's word for it that Schouw, somewhere in this impenetrable text, writes, with inter-letter spaces for extra emphasis: "Plants that occur near cities and towns are called *P l a n t a e U r b a n a e*; for example *Onopordon Acanthium* [cotton thistle], *Xanthium strumarium* [common cocklebur]. In most cases, an alien origin is the cause by which these plants are found only in the vicinity of cities and towns."

It's interesting that two centuries ago, botanists were already recognizing how exotic species form an important contribution to the urban biodiversity. In those days, the major avenues that today are bringing exotic plant species into cities did not yet exist: there were no garden centers, scattered bird seed, or globalized agricultural produce. And the pet trade, as well as the planes, trains, and automobiles on which far-flung fauna nowadays hitch accidental rides, were also not yet as commonplace as they are now. With these activities bringing so many exotic species into cities, it is no surprise that nowadays, much more strongly than in Schouw's time, urban biodiversity is an eclectic mix of species from all over the world. In European and North American cities, the wild flora consists of 35 to 40 percent exotic species. And in the city center of Beijing, this figure even is 53 percent. Sometimes, the role that socio-economic factors play in these patterns is all too apparent. In Phoenix, Arizona, botanists measured the plant diversity in more than two hundred plots, each 30 by 30 yards, strewn randomly across the city and its surroundings. They discovered that one of the factors most strongly determining how many types of plant they would find in a plot was the affluence of the local neighborhood. The better-off the residents, the greater the diversity of plants. This effect (which they called the "luxury effect") is a clear sign that travel and trade, and the relentless escapism of exotic plants from well-tended gardens, are responsible for the botanical enrichment of urban centers.

These constantly arriving foreign inhabitants form the first of at least four explanations for the high biodiversity urban naturalists are encountering in their cities. A second explanation is the fact that the places where people like to build their settlements, which then grow into cities, are often biologically rich areas to begin with. If you open an atlas and look where the world's biggest cities are, you'll notice that they are not on mountain plateaus, deserts or other biologically poor regions. Instead, they are in the exact same places where we find biodiversity hotspots: estuaries, flood plains, fertile low-lying areas, and other places where humans and wildlife alike find plenty of food and many different niches are available. In other words, the second reason for the rich biodiversity of a city is that it was already rich before the city was built. Some of this richness will have clung on in the remaining patches of habitat, embedded in the city as it develops. As we saw in an earlier chapter, much of Singapore's native flora and fauna lives in those slivers of primary forest that have survived amid the city's growth.

A third source of urban biological richness is, in fact, the loss of good-quality habitat immediately outside the city perimeter. These days, many urban centers are ecological oases compared with the surrounding countryside. In the past, it was the countryside (with its romantic small-scale hodgepodge of fields and pastures, hedges and bushes, brooks and ponds) where the landscape was richly varied and every nook and cranny provided habitat for different species. That is why New York–based naturalists in the nineteenth century would venture out into the *vicinity* of the Big Apple. Compared with such pastoral bliss, the biodiversity of the beat-up wasteland of the inner city, with its factories and pollution, was decidedly poor. Today, in many countries, the tables are turned. In the agricultural countryside, little or no space for biodiversity is left among the obsessively manicured fields and plantations, bisected by machine-dug canals straight as a die, where maximum agricultural production is squeezed out of every square inch of land—especially as expanding cities eat up more and more arable land. Compared to such sterile, geometric landscapes, the messiness of the urban center, a varied mix of backyards, green roofs, old stone walls, overgrown drains, and city parks, is a haven for a lot of wildlife.

Botanists Zdena Chocholoušková and Petr Pyšek documented this turning of the tables for the city of Plzeň in the Czech Republic, for example. They worked their way through piles of old publications, reports, and herbariums to record the changes in the flora of the city and its surroundings over the past 130 years. Inside the city, they saw that the number of plant species rose steadily, from 478 in the late nineteenth century to 595 in the 1960s and 773 today. In the surrounding countryside, by contrast, the trend was opposite: from 1,112 to 768 to 745. Why? Probably because in the twentieth century, the countryside, with its increasing agricultural intensification, had become more hostile to plant life, while in the city center, the reverse had happened. With a little poetic license, you could say that weeds, prosecuted and outlawed in the country, had taken up refuge inside the city walls.

Organisms that almost literally enjoy sanctuary in cities are large vertebrate animals. Brush-turkeys in Sydney, coyotes in Chicago, foxes in London, leopards in Mumbai, and mugger crocodiles in Gujarat . . . All over the world, cities have witnessed an influx of large, and often dangerous, birds, mammals, and reptiles. Of course, given their size, sometimes these are simply the eye-catching tip of the iceberg of thousands of similar, but less visible, changes in the urban biodiversity. But in the case of this megafauna, often it's the relaxed attitude of urbanites that makes the city more welcoming than their native habitat.

Take coyotes, for example. Ever since the *American Midland Naturalist* in 1980 published an article on the behavior of a renegade coyote living in downtown Lincoln, Nebraska, these canines have been moving into cities hand over fist. Stanley Gehrt, an urban zoologist at Ohio State University, has been tagging hundreds of coyotes in Chicago with ear tags and microchips. He estimates that there are now more than 2,000 in the city. Four hundred of those he has fitted with radio and GPS collars and followed them around on their prowls along railway lines, watched them waiting at traffic lights and raising their young on the roof of a parking garage. Jaded city slickers as they may be, one of the main benefits they enjoy in the city seems to be the lack of persecution. Compared to rural coyotes, the ones in cities have a fourfold lower chance of dying a violent death. "We're now seeing generations of certain carnivores that had had fairly light amounts of persecution by people," Gehrt told the magazine *Popular Science* in 2012. "They may view cities quite a bit differently than their ancestors did 50 years ago. Then, if they saw a human, there was a good chance they were going to get shot."

"Same thing here, mate," the Australian brush-turkey (*Alectura lathami*) might have butted in from the other side of the globe. For centuries, this species of "incubator bird" (fowl that build huge mounds of sand and leaves to let the heat of rot incubate their eggs) was a favorite snack to shoot and fry when you were out in the bush. Today, thanks to a hunting ban imposed in the early 1970s, this former bush tucker has made a spectacular comeback. Not only in rural Australia, but, unexpectedly, especially in cities, where the prohibition is presumably respected better than in the outback. Brush-turkey expert Darryl Jones of Griffiths University says that, in the past twenty years, the Brisbane population has grown sevenfold, with Sydney set to surrender next. That this particular large bird would ever develop an urban streak was unexpected, since its nesting habit would seem to be impossible to maintain in cities. Not so: the birds simply dig up people's backyards and use entire garden beds to construct their nest mounds that can weigh up to four tons. (Now there's an ecosystem engineer if ever there was one!) Small wonder the Australian Broadcasting Corporation advises people to do the following to minimize brush-turkey damage: (1) place rocks around precious plants, and (2) attract the bird to a less valuable area of your garden by building a compost mound yourself. Crikey, that's almost as much work as replanting your flower bed!

The fourth and final explanation for the rich biodiversity of cities is the sheer diversity of habitat patches. Think about it: when we view a city with human eyes we may see shopping streets, parking lots, thoroughfares, business districts, and pedestrian zones. But to a peregrine falcon soaring overhead, a hover fly cruising along a main street, or a fluffy milkweed seed parachuting down, the city is a kaleidoscope of rocky ledges, humid pits, strips of moss, and underground streams. These scattered bits of habitat form a surprisingly varied landscape, with a multitude of niches jointly supporting a rich, but heavily fragmented, biodiversity.

Consider the endless variety that exists among urban gardens: sterile ones smothered in tiles, pebbles, and perfectly groomed exotic bushes . . . vertical green walls . . . messy backyards that nobody pays any mind . . . gardens composed of nothing but a fenced lawn . . . roof gardens with potted palms and rock-dwelling herbs . . . vegetable gardens . . . swampy gardens with a pond and a slippery rocky slope . . . In our age of individuality, there are as many types of garden as there are gardeners. In 1999, a team of biologists from the University of Sheffield, led by the prolific ecologist Kevin Gaston (he has since moved to the University of Exeter), began a multi-year project to study the ecology of urban gardens in Sheffield. The name of the original project is BUGS: Biodiversity of Urban Gardens in Sheffield.

To begin with, the BUGS team conducted a telephone survey. Choosing random entries from the city's phone book, they posed a set of garden-related questions to whomever picked up. That is, provided the homeowner was amenable to collaboration, because, as the team dryly write in one of the scientific papers, "in some cases, the call was terminated before its purpose could be conveyed to the recipient." Based on 250 interviewed homeowners, they estimated that the 175,000 domestic gardens in this city of 500,000 inhabitants, jointly covering about a quarter of the entire city surface, hold 25,200 ponds, 45,500 nest boxes, 50,750 compost heaps, and 360,000 trees. In other words: a huge ecological resource. Yet, urban gardens are rarely included in any account of a region's green spaces. And far from being the biological deserts that ecologist Charles Elton (in his 1966 book *The Pattern of Animal Communities*) proclaimed they were, the BUGS project proved them to be replete with wildlife.

The BUGS people found sixty-one homeowners willing to allow a rather overwhelming invasion of their privacy. Anyone who has ever watched a group of field biologists given free rein knows what this means. The team took out measuring tapes to obtain the exact dimensions of each garden and its types of ground cover, and drew a sketch map on the spot. They went about with plant guides and notebooks and identified each and every tree, bush, and herb that they could find, including those in pots and ponds. While they were at it, they also collected leaves with insect "mines": those squiggly tunnels made by the larvae of certain moths, flies, or other insects. Most of these mines are so characteristic that an expert can tell which species made them without even seeing the actual animal.

Along the edge of each garden, they placed three "pitfall traps":

white plastic coffee cups dug into the ground, aimed at trapping insects and other arthropods that blunder into them and cannot get out. To make sure, the team filled the cups with alcohol. Normally, a more toxic chemical, ethylene glycol, is used for this, but, as the team wrote, ethanol was used instead because of "the risk of being found by pets or children." They then dug up bags full of dead leaves and soil to look for other invertebrates, and, as if all that weren't enough, they also built a Malaise trap to catch flying insects (the same kind of contraption that Denis Owen used in his Leicester garden, as we saw in the previous chapter). Despite their scorched-earth approach to urban garden ecology, the team reports that in most gardens they still were offered tea and biscuits.

In these sixty-one garden-size field sites they found 1,166 different plant species. As is to be expected from planted gardens, the majority of those species (70 percent) were exotic. But still, 344 species (that's a quarter of the entire British flora!) were native species. The 30,000 or so invertebrates they found belonged to roughly 800 species. No numbers to be sniffed at, but also not huge, compared with what Denis Owen found in a *single* garden. But what is more important is not the sheer numbers of species, but rather the changes from one garden to the next. About half of all the species of insects and spiders that they found were living in just a single garden. And when the team made a so-called "accumulation curve," which showed how the overall tally increased with every new garden added to the list, the curve showed no signs of leveling off. In other words, every garden has an almost completely different flora and fauna.

So that is just sixty-one gardens, a vanishingly small fraction of all Sheffield's gardens, which, in turn, is just a sliver of the entire garden area in the whole of the UK. Just imagine the combined biodiversity that all those yards, gardens, and plots hold.

And imagine the same for all those other forgotten bits of habitat in the city: neglected gutters, roadside verges, mossy roofs . . .

Of course cities are challenging in their own way, but at least for animals and plants that can disperse over long distances and are able to survive in those small, isolated pockets of habitat, the city is a surprisingly varied, mosaic landscape, offering more microhabitats to a great array of species. Add to that the other three reasons for urban biological richness (exotic species, pre-existing biodiversity hotspots, sanctuary from persecution), and we begin to understand why the species lists that the urban naturalists of the previous chapter are compiling are so lengthy.

Lengthy, perhaps, but not random. Not every kind of animal or plant can live its life in the city. Some species have properties that make them less suited for life as a city slicker, whereas other species are cut out for the job. Those properties, and how they evolve, form the core of this book. So let's move a little bit closer to that core by looking at the fascinating phenomenon of urban "pre-adaptation."

6
IF I CAN MAKE IT THERE

WE ARE PACING IN FRONT OF THE
central station in Leiden, my hometown. Like most
Dutch railway stations, it overlooks an ocean of parked bicycles.
A two-story open-air bicycle parking garage stretches on either
side of the main entrance, and is crammed with bikes, their
thousands of chrome handlebars glistening in the morning sun
like ripples on a calm inland sea. While I can survey this tangle of metal spokes, springs, tubes, frames, cogwheels, and
chains, my companion, famed biologist Geerat Vermeij of the
University of California at Davis (and a regular visitor), cannot.
Blind since the age of three, Vermeij has built a career as a pal-

aeontologist, ecologist, evolutionary biologist, and best-selling author using his other delicate faculties of fingertips, sensitive ears, and powerful brain.

That Vermeij can still detect the mass of bicycles in front of us is thanks to its being the habitat of one of the most endearing, yet underappreciated, urban birds: the house sparrow (*Passer domesticus*). Hopping about on the ground among the wheels, taking dust baths in the sandy patches between tiles, perched on spokes and flitting from saddle to rear carrier, the troupes of the brownish-gray birds are creating a continuous rustle of wing flutters and chatty chirps. Vermeij smiles, and sparkles of tender creases appear in the corners of his unseeing eyes. "Yes," he says. "You are right: they are everywhere."

I brought Vermeij to this bicycle-rack-inhabiting population of sparrows to discuss the power of pre-adaptation.

Pre-adaptation is a bit of a mysterious and controversial term in evolutionary biology. After all, evolution is nature's hindsight: today's adaptations are caused by yesterday's natural selection. So how can an animal or plant be "pre-adapted," when evolution cannot look into the future and cannot prepare any organism for things to come?

Let's go back to those sparrows. The habitat they are occupying in front of Leiden's train station is not a habitat the species ever evolved to occupy. In its evolutionary past, it never encountered any bicycle racks. And yet, as Vermeij and I observe these birds with our ears and eyes, respectively, they seem perfectly suited for a life among the spokes. Their short wings are ideal for taking brief flights from pedal to saddle. They scurry about in groups and in the dense tangle of metal keep in touch with one another with continuous brief chirps. They have a habit of taking off en masse and then, at the least cause for alarm, dissolve among the parked bikes. The reason they seem so at home here probably is that the house sparrow's natural habitat is thickets of thorny trees and bushes. To them, the vast expanse of metal rods of varying thickness, density, inclination, and curvature perfectly resembles the brushwood of their original home.

Not that we really know what that natural home is. Like the house crow, the house sparrow is one of those birds whose life has become so entwined with human habitation, that it no longer occurs in the wild. Its ancestors in pre-human times probably were specialists of half-open bush-land in dry areas, groupwise nesting in shrubs, feeding on seeds and insects, and retreating into their spiny shelter whenever a sparrowhawk would show itself on the horizon. Then humans and their agriculture appeared and the house sparrow became one of those species that deserted its natural habitat in favor of human presence, feeding on discarded grains, shacking up with them in roofs of houses and stables. And, eventually, bicycle racks.

In other words, *Passer domesticus* has become an urban species because it was already adapted to a lifestyle that, purely by accident, prepared it for the niches that we have created in cities. The urban environment offers conditions that happen to resemble one or more aspects of a species' way of living in pre-urban times. And it is those species that are pre-adapted to the novel niches in the city. They are the first to move in.

Besides house sparrows, there are other pre-adapted birds that live in and around the Leiden train station. The city pigeons perched over the grand clock at the main entrance are descendants of the wild rock pigeon (*Columba livia*), a species native to Europe and North Africa, where it naturally occurs only in areas with rocky cliffs—for roosting and nesting. Obviously, the swampy lowlands of the Netherlands, devoid of any accidentation higher than a mole hill, let alone cliffs, were never part of its natural range. Until, that is, humans began erecting artificial cliff faces: brick and concrete buildings with ledges and window-sills ideal for these birds to perch on—even if we try to discourage that with rows of plastic needles.

Likewise, the sooty, sickle-winged swifts (*Apus apus*) that, screeching, are criss-crossing the sky above us, are also typical cliff-hangers. They have made Leiden their own thanks to the gaps under the zinc gutters in 1970s housing estates, the openings beneath the roof tiles of the seventeenth-century church and between the bricks of the old windmill—ideal nesting sites for these birds of rocky terrain. The black-and-white oystercatchers (*Haematopus ostralegus*), with their long bright red bills that are pacing across the lawns behind the station as if they were at home, are originally coastal birds, nesting on beaches and using their strong bills to draw clams from the mud. In Leiden, they have traded mudflats for lawns, clams for earthworms, and pebbly beaches for the flat roofs of the Leiden University Medical Center. Each of these bird species—the house sparrows, rock doves, swifts, and oystercatchers—was in one way or another pre-adapted to life in the city. They were the chosen ones that the urban environment has selected from all available avifauna.

It's fairly obvious why the birds flying around the Leiden central station may be pre-adapted (or predisposed, as Vermeij prefers to call it—to dispel the false notion of evolution thinking ahead). The link between the character of their original habitat and that of the inner city is plain to see. In a similar vein, many of the tiny arthropods that live in our houses are species that originally dwelled in caves—and some may even have been moving buddies of our ancestors as they decided to stop being cavemen and become housemen instead. The closest relatives of bed bugs (*Cimex lectularius*) are parasites of cave bats—which indicates that that was also bed bugs' original niche. The cellar spider (*Pholcus phalangioides*), which lives in houses all over world, simply likes stony, dank, enclosed spaces. It naturally occurs in caves and caverns; the hollow spaces we occupy inside brick and concrete shells ("houses") are, to this spider, no better or worse than its underground habitat in the wild.

But some other pre-adaptations are a bit more elusive. The relentless "predation" by traffic, for example. "We can see and hear traffic approaching," says Vermeij, "but roadkill shows that some birds cannot. And why do some bird species fly into glass windows, but others don't? I find that an interesting question. Also, some kinds of bird, like crows, seem to be global champions at surviving in cities and suburbs. And why is the American Robin an urban bird in North America, but none of the other members of its family?"

One way to get a handle on such less than obvious pre-adaptation is by looking for common patterns across different cities. Ecologists Carmen Paz Silva and Olga Barbosa of the Universidad Austral de Chile, for example, used such an approach in three medium-sized cities in Southern Chile: Temuco, Valdivia, and Osorno. These cities (each with 100,000 to 350,000 human inhabitants) lie in an area rich in biodiversity, the so-called Valdivian Rain Forest Ecoregion. Silva, Barbosa, and their colleagues first drew a grid of 250 by 250 yards mesh width over each city and its surroundings. Then, they randomly selected 110 of those grid cells in each city, and 50 in the rural vicinity of each city (so, 480 cells in all), and went to each to see which birds occurred there. They did this simply by picking a spot in the center of the grid cell, and standing there for six minutes in the morning time to record all the birds seen or heard during that time.

This work, which spanned the whole of the 2012 breeding season, showed that the birds inside these cities were not a random selection from the avifauna in the surrounding countryside. In each city, a similar set of urban birds dominated, composed of species like the Chilean swallow (*Tachycineta meyeni*) and the Chimango caracara (*Milvago chimango*), as well as, again, the cosmopolitan house sparrow and rock pigeon. Conversely, birds such as the robin-like chucao tapaculo (*Scelorchilus rubecula*), though very common in the Chilean countryside, would never venture into town, and neither would the fire-eyed diucon (*Xolmis pyrope*) and the stunning Patagonian sierra-finch (*Phrygilus patagonicus*). Still, for many birds the differences were not that stark: they would occur both in and outside the city, but at slightly different proportions.

To figure out what were the crucial urban pre-adaptations, Silva and Barbosa first classed each bird species as either carrion, fruit, seed, insect, or nectar feeding, carnivore, or omnivore. Then, they recorded what their natural habitat was: forest, open terrain, water/wetland, or ubiquitous. Finally, they performed a series of statistical tests to see what the urban bird-life should look like if a random selection had been made from the available bird types, either regarding habitat or food preferences. This analysis told them that the urban bird composition was anything but random. For urban birds in Southern Chilean cities, it definitely helps to be an omnivore or a seed-eater, and to be not too picky about the habitat you live in. This makes sense: as we saw in the previous chapter, cities are mosaics of different types of habitats, so a species that, outside of cities, is adapted to variable, unpredictable environments (for example, dynamic, unstable

places like forest gaps or flood plains) is well honed for the life of
the city. But why seed-eaters? This is because humans are also
primarily seed-eaters. Much of our diet is based on grains, so most
of our food-scraps (crusts of bread, discarded cooked rice from the
bottom of the pan, but also half-eaten crackers and crumbs of
biscuits) fit perfectly into the diet of birds that normally eat seeds
and nuts.

So city birds include those that like rocky or tangled substra-
tum, and ones that either share our food preferences or are not
too picky about their living quarters. But there is more to being
a city bird than food and flexibility. Think of communication:
most birds communicate with one another via sound. How
to do that against the combined din of traffic, sirens, alarms,
shouting people, and power tools? Silva and Barbosa's colleague Clinton Francis at the University of Colorado set out to
see which birds are better able to deal with such human noises.
Surprisingly, to study this he did not set up an experiment in a
big city. Instead, he headed for the desert of northern New
Mexico.

Here, in the deserted Rattlesnake Canyon, there is no urban
development to speak of. But there is human-generated noise.
The area, one of the nation's most productive fossil fuel sites, is
dotted with some 20,000 oil and gas wells. Some of those gas
wells are equipped with noisy compressors: pumps that force
gas out of the ground and into pipes day and night. Other gas
wells make do without a compressor and are blissfully quiet.
Francis realized that here was an ideal "natural laboratory" to
study the effect of noise on birds, without any of the problems
of comparing city and countryside. After all, in cities, noise is
accompanied by all of the other changes in the environment
that we have come across before. So if we find that, say, mock-

ingbirds are less common in cities than outside of cities, there is no way to tell for sure that this is due to noise. It could be due to any of the other factors that characterize the urban habitat. But out in the desert, the environment was piñon-juniper woodland all around, either with or without a painfully loud compressor groaning away in the background. An experimenter's dreamland.

What Francis and his team did was similar to Silva and Barbosa's study in that they picked a set of noisy and a set of silent gas wells and spent seven minutes looking and listening for birds. At the noisy well pads, they persuaded the oil company's managers to turn off the compressor for that duration, because the noise would also have hampered their attempts at detecting the birds. The results were clear-cut: birds with low-pitched calls and songs, such as the mourning dove (*Zenaida macroura*), were absent from sites with compressors. The sound of the machines was such that these birds simply could not make themselves heard any more and therefore had left. Birds with high-pitched voices, on the other hand, did not seem to care where they lived: the soprano calls and songs of species like the chipping sparrow (*Spizella passerina*) carried well above the baritone of the gas pumps. There were even some birds, for example the black-chinned hummingbird (*Archilochus alexandri*), that actually preferred to build their nests near the compressors—the closer the better. Francis thinks that this is because their predators, Woodhouse's scrub jays (*Aphelacoma woodhouseii*), cannot tolerate the noise. So for the hummingbirds, the noise actually affords protection.

Surprise, surprise: in cities, where the noise is also mostly low-frequency, the commonest birds seem to be the ones with relatively high-pitched voices. But it took research out in the desert to prove the link between noise-pollution and pre-adaptation of the Mariah Careys among birds.

So pre-adaptation is crucial for urban ecosystems. It determines which species are filtered by the sieves of concrete and cars, garbage and grime, and get to hang their hats on the city street signs. The urban flora and fauna are largely made up of native and exotic species that, coincidentally, have evolved to deal with one or more challenges similar to the ones posed by cities.

Let's go back to myrmecophiles for a minute; those animals that have evolved to live inside ant societies—we met them at the beginning of this book. They, too, are not a random selection of insects and other invertebrates. In an article in the journal *Myrmecological News*, Joe Parker of the California Institute of Technology claims that pre-adaptation lies at the root of the compendium of myrmecophily as well. Many of them are clown beetles, he says. These beetles have sturdy wing covers, which make them resemble armored vehicles, and this protects them against ant attack. It has made them succeed in penetrating ants' nests where lesser insects have failed. Similarly, a lot of myrmecophiles belong to the pselaphine rove beetles, which have internal reinforcements of their body, allowing them to be bitten and squeezed by an irate ant without much ill effect. And some other myrmecophiles are aleocharine rove beetles, which all have a gland at their back-end that helps them to wage chemical warfare on ants.

So, what is happening in our cities is perhaps similar to what happened millions of years ago, when small soil animals had the audacity to penetrate the first ant colonies. Those species that were pre-adapted to the toils of life inside an ant nest were the ones that evolution then further improved and molded into expert myrmecophiles. Compared with the long evolution of myrmecophiles in ant societies, pre-adapted animals and plants have only just begun setting themselves up in human cities. But that does not mean that the initial stages of further evolutionary improvement of their city streak may not be going on.

II.

CITYSCAPES

We see nothing of these slow changes in progress, until the hand of time has marked the long lapse of ages.

CHARLES DARWIN, *On the Origin of Species* (1859)

I don't think so.

HOMEY D. CLOWN, *In Living Color* (1990)

7
THESE ARE THE FACTS

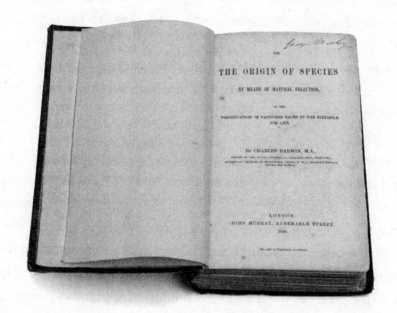

ALBERT BRYDGES FARN WAS BORN in 1841. His entry in *The Aurelian Legacy*, a sort of Who's Who for British collectors of Lepidoptera (butterflies and moths), describes him as "an all-round naturalist," "a man of vigor, courage, and rather boisterous good humor." It also calls him "a sportsman," which in those days did not imply he would be seen jogging along country lanes or playing rugby with the village lads, but rather referred to his reputation for potting at bats with a .22 rifle and having bagged a legendary thirty snipe with thirty consecutive shots on Lord Walsingham's Estate. Farn, clearly, liked to kill stuff.

Most of what he killed, though, were butterflies and moths, which he pinned, mounted, labeled, identified, and organized with great precision. When, in 1921, he died, he left what many considered the finest private lepidopterological collection in Britain at the time. Sadly, the collection was auctioned off piecemeal, and bits of it ended up in different places. Some, says Adam Hart of the University of Gloucestershire, are now "in the bowels of the Natural History Museum" in London. Though he does not know for sure, Hart likes to think that among them are some specimens of the Annulet moth (*Charissa obscurata*), which Farn collected near Lewes in the 1870s.

It's a rather drab species, the Annulet, paling in comparison to some of Farn's prize possessions, like the spectacular Purple Emperor (*Apatura iris*) that he caught in South Wales. Or the rows and rows of pinned European Map butterfly (*Araschnia levana*), a pretty black-orange-white species from continental Europe, illegally introduced into the Forest of Dean in 1912 only to be single-handedly exterminated by Farn, who disapproved of all exotic species, pretty or not. Despite its humdrum appearance, though, the Annulet is Farn's claim to fame. A fame, however, that came 130 years late.

In 2009, Hart, a professor of science communication, was visiting the Gloucester City Museum and Art Gallery in preparation for a class. "I was going through some backroom stuff to look for teaching specimens," he says. There, he came across a print-out of a letter dated November 18, 1878. The letter had been written by Farn, and the print-out was there because the museum owns a copy of an annotated book that had once belonged to Farn, and the librarian had taken an interest in him. The reason that this particular letter survives and had even been transcribed and put online is, however, not because of the author, but because of the addressee: Charles Darwin.

By 1878, the aging Darwin was one of the most famous scientists in England. A new generation had grown up since *On the Origin of Species* had been published, and his name as Mr. Evolution was firmly established. Colleagues from all over the world corresponded with him and Darwin kept a meticulous administration of the letters he received and sent—not for social but for scientific reasons. What his correspondents conveyed to him was crucial for his work. As the archivists of the Darwin Correspondence Project at Cambridge University (where much of Darwin's library is kept) evocatively explain, "[h]e went back over some letters again and again as he worked on different subjects, scrawling on them in different colored pencil, and cut them up so that he could file the pieces with relevant notes or stick them into his experiment book. Letters were dissected like specimens, every useful bit of information sucked out of them and then reincarnated in his publications."

As far as we know, Albert Farn wrote to Darwin only once. The letter survived in Darwin's library, the Darwin Correspondence Project duly transcribed it, placed the text online and it was a print-out of this text that Hart found lying around. It is only a brief note, and it seems that Darwin never did anything with it or replied to it.

Farn writes:

My dear Sir,
The belief that I am about to relate something which may be of interest to you, must be my excuse for troubling you with a letter.

Perhaps among the whole of the British Lepidoptera, no species varies more, according to the locality in which it is found, than does [the Annulet moth]. They are almost black on the New Forest peat; gray on limestone; almost white on the chalk

near Lewes; and brown on clay, and on the red soil of Here-
fordshire.

Do these variations point to the "survival of the fittest"? I
think so.

It was, therefore with some surprise that I took specimens as
dark as any of those in the New Forest on a chalk slope; and I
have pondered for a solution. Can this be it?

It is a curious fact, in connexion with these dark specimens, that
for the last quarter of a century the chalk slope, on which they
occur, has been swept by volumes of black smoke from some lime-
kilns situated at the bottom: the herbage, although growing
luxuriantly, is blackened by it.

I am told, too, that the very light specimens are now much
less common at Lewes than formerly, and that, for some few
years, lime-kilns have been in use there.

These are the facts I desire to bring to your notice.

I am, Dear Sir, Yours very faithfully,

A. B. Farn

"I must admit it was a bit of a eureka moment," says Hart. "This letter had been lying around for so long but no one had realized its significance!" That significance, as Hart pointed out to the evolutionary biology community in a 2010 article in *Current Biology*, was of course that Farn's observation may have been the first recorded case of ongoing natural selection. What Farn was suggesting was that the light-colored Annulets, originally well camouflaged on the pale limestone, had now become sitting ducks against the soot-blackened background, and were being picked off by birds and other predators. Meanwhile, a genetic mutant with dark wings had appeared and had been "naturally selected" because it did not stand out as much as its pale ancestors. If Farn was correct, it would be the very first observation of evolution in action. As Farn rightly anticipated, Darwin should have been thrilled. So why did he ignore Farn's letter?

Of course, it is possible that Darwin just couldn't be bothered on that November 18, 1878. Maybe he was tending to his orchids, or playing with his grandchildren, or laid up with one of his fits of general malaise. But of course we prefer to see a deeper significance in his lack of response. If it means anything, my guess would be that Darwin underestimated the power of his own discovery, natural selection, and that he found it hard to imagine that its work could be observed on the timescale of years or decades. After all, in Chapter IV of *On the Origin of Species*, he wrote, "We see nothing of these slow changes in progress, until the hand of time has marked the long lapse of ages."

In the preceding pages of his great book, Darwin had laid out the foundations of his theory in four easy, steadfast steps. One—there is variation: in many (sometimes near-imperceptible) ways, each individual is different from the next one. Two—this variation is heritable: offspring resemble their parents. Three—there is surplus: most offspring do not survive. Four—there is selection: survival is not random but favors those who are best suited to the world they live in. To Darwin's mind—and to everyone since who has fully grasped the enormity of this insight—natural selection is a law of nature. As Darwin wrote, "natural selection is daily and hourly scrutinizing, throughout the world, every variation, even the slightest; rejecting that which is bad, preserving and adding up all that is good."

And yet, despite that "daily and hourly," Darwin did not actually believe that natural selection could be observed in real time. Maybe this was because he lacked the mathematical prowess to calculate exactly how long it would take for natural selection to do its thing. It lasted until the 1920s for mathematical biologists like J.B.S. Haldane and Ronald Fisher to do this. With Darwin's theory cast in algebraic formulas, it became possible to see whether his pessimism was well founded or not.

As it turned out, it was not. Darwin's mistake probably was that he imagined natural selection to be a linear process. He may have thought to himself, let's imagine a population of 100,000 pale-winged moths. Then, a black-winged mutant appears that enjoys a teeny-weeny advantage. Say, an imperceptible 1 percent—meaning that for every 100 black-winged moths born, surviving and reproducing, there would be 99 pale-winged ones. It's that small a difference. So how long would it take for all those 100,000 white-wings with one black-winged mutant thrown in, to evolve into a fully black-winged moth population and all pale-winged gone? Forever, right? Wrong—it takes only about a few hundred generations.

That is because natural selection is not a linear process. In the beginning, when the black-wings are still rare, they increase only very slowly, one moth at a time. But when the frequency of black-wings has gone up to a few percent, the process speeds up, because all those thousands of black-winged moths enjoy the same advantage, and pour their joint offspring into the total gene pool, which becomes duskier by the day.

You can see this for yourself by doing an online simulation. Radford University, for example, has a website where you can enter the population size, the advantage of a mutant (the so-called selection coefficient), and the start-off frequency of the mutant, and you can just watch a virtual population evolve in a nice, S-shaped curve. Play around with the settings and you'll see that it doesn't make much difference whether your moth population is 10,000 or 100,000 or even 1 million winged souls: in all cases they evolve into a black-winged moth in less than a thousand generations, while enjoying only a 1 percent advantage. Make that selection coefficient 5 percent and it all happens in just 200 generations. For some moth species, 200 generations is less than a century. So, at least in theory, even very weak natural selection can have dramatic effects before the hand of time has begun to mark any longish lapses.

It seems Darwin never really entertained the notion that such evolutionary agility was a possibility. Although . . . In the first four editions of *On the Origin of Species*, he still writes emphatically, "I do believe that natural selection will always act very slowly." But in the fifth edition, published ten years after the first, he changed "always" for "generally," so he may have begun to doubt that natural selection is so tardy a process. Be that as it may, Darwin missed a trick by not picking up on Farn's tip, and it was left to the next generation to reveal the breakneck evolution of "industrial melanism." Not in the Annulet moth, but in *Biston betularia*. The "peppered moth," as it is called, is literally a textbook example of urban evolution, and you've probably heard about it in school. But there have been so many recent twists and turns to the story that I hope you'll forgive me for re-telling it.

8
URBAN MYTHS

WE MAY THINK OF RAPID URBAN growth as a thing of today, but between 1770 and 1850, the city of Manchester grew as explosively as any twenty-first-century megalopolis: from 24,000 to 350,000 inhabitants. The city's coal-powered textile industry sucked in workers from the surrounding countryside and spewed pollution into it. Immense quantities of soot, sulfur, and nitrous gases billowed from its smoke stacks, which darkened the sky, blocked out the sun, and on windless days caused a haze so thick that people could barely make out their neighbors across the street. A constant mist of soot particles settled on everything: the houses, the pavements, even the trees in the rural environs of the city.

Imagine an autumn day in 1819. In a forest just outside Manchester, something like this must have happened. A caterpillar of the peppered moth (*Biston betularia*) is inching down the stem of a soot-blackened birch tree, heading for the ground to begin its pupation. As proper inchworms do, it clamps the bark with its true legs (at the front), then pulls its squishy false legs (at the rear end of its long stick-like body) up to its regular legs, and folds its body into the shape of an omega. Then, it releases its true legs, holding on to the bark with only its false legs, and stretches forward, brings up the rear, does the omega, stretches, brings up, releases, grabs, pulls up . . . tirelessly until it reaches the foot of the tree.

Although the caterpillar is, by definition, immature, its testes are already formed inside and busy creating sperm cells for when, after pupation, the animal will metamorphose into a sexually active adult peppered moth, its wings characterized by a sprinkling of black marks on a white background. Or at least, that is how its parents, and all the other peppered moths in Britain until that day, looked. But as our caterpillar takes a final bound onto the grass beneath its tree, something odd happens in one testicular cell. Something that is to change the course of peppered moth evolution. While the cellular machinery is separating the chromosomes and packaging them into what is going to be a sperm cell, a bit of DNA releases itself from one of the chromosomes. It is a so-called transposon, a "jumping gene," able to cut itself out of a chromosome and reinsert itself somewhere else. And that is precisely what this transposon does. Unbeknown to the caterpillar, who is busy pushing its head between the grass roots, so-called transposase enzymes cut the 22,000-letter-long runt of

DNA loose from its original position and insert it smack in the middle of *cortex*, a gene controlling wing pigmentation in moths.

While the caterpillar burrows into the soil, turns itself into a pupa, hibernates and finally hatches into a moth, the mutated sperm cell sits quietly, biding its time. It obediently joins thousands of other, non-mutant, sperm cells in one of the moth's sperm packages, is ejaculated into a female peppered moth upon one of the male moth's successful copulations, and, by sheer luck, manages to fertilize one of her eggs. The fertilized egg develops into a young caterpillar, built from cells that now all carry a copy of the mutated *cortex* gene. For that whole summer, the mutant caterpillar, together with all its siblings, munches away at birch foliage until it is its time again to burrow into the ground and pupate.

But as the chrysalis lies there quietly beneath the grass roots, its dormant appearance belies a revolution going on inside. In the animal's developing wings, still encased inside the auburn shell of the pupa, the transposon wedged into the *cortex* gene proves to be a spanner in the works that normally produce that delicate white-and-black speckled wing pattern. Instead, when the moth emerges, scrambles up, and clings to a branch of birch, its unfolding and hardening wings turn out to be a pure anthracite black. Not unlike the tinge of the soot clinging to that very same branch.

Our black-winged *Biston betularia* survives and reproduces. It spawns a small but slowly growing band of black descendants. Some of these are noticed by early nineteenth-century Mancunian entomologists, but the first catch that makes it into the scientific literature is one that is netted and pinned in 1848 by Manchester moth collector R.S. Edleston. From then on, things escalate. The black moths proliferate. In the 1860s, in some parts of Manchester, the black mutants are becoming commoner than the pale ones. From their Manchester stronghold, the black gene also filters into other parts of England. In the 1870s, black moths are seen in Staffordshire, some 40 miles south of Manchester, and in Yorkshire to the northeast. By the late nineteenth century, the original gene for pale wings had almost gone from many of the British populations of *Biston betularia*, some parts of the rural south excepted. The continent and North America succumbed not much later.

Moth boffins are baffled, and several debates are raging in the British entomological journals, with speculation ranging from changes in humidity and food to "the powerful impression of surrounding objects on the female" during procreation (these are the days before genes and how they work are discovered). But it is the eminent Victorian lepidopterologist J.W. Tutt who, in his 1896 book *British Moths*, eloquently promulgated the concept of what we now know as industrial melanism:

> Let us see whether we can understand how this has been brought about! [. . .] In our woods in the south the trunks are pale and the moth has a fair chance of escape, but put the Peppered Moth with its white ground color on a black tree trunk, and what would happen? It would, as you say, be very conspicuous, and would fall prey to the first bird that spied it out. But some of these Peppered Moths have more black about them than others, and you can easily understand that the blacker they are the nearer they will be to the color of the tree trunk, and the greater will become the difficulty of detecting them. So it really is; the paler ones the birds eat, the darker ones escape.

"So it really is." Most evolutionary biologists today would agree with Tutt's explicit (and Albert Farn's implicit) explanation for industrial melanism. Acid rain killed the lichens, soot turned the naked branches black, and the mottled coloration of the peppered moth, the Annulet, and many other insect species was rendered ineffective as camouflage. New or existing mutants with darker bodies, which previously would never gain a foothold, now proved better at not being seen against the darkening background. And natural selection did the rest.

But the general acceptance of this process has had a history as peppered as *Biston betularia's* wings. After all, more was required than just the regarded opinion of an esteemed lepidopterist. For Tutt's just-so story to be adopted as the first rubber-stamped case of real-time evolution, proof and verification were needed.

First to take a stab at the peppered moth was the mathematical biologist J.B.S. Haldane. In 1924, he used the time (50 years) it had taken for the dark moths to take over Manchester to calculate the selection coefficient, that is, the relative disadvantage of the pale moths compared with the dark ones. He came up with a value of about 50 percent, meaning that for every two pale moths surviving bird attacks and breeding, there were three dark ones. At the time, many of Haldane's colleagues balked at the suggestion that selection could ever be that strong. Moreover, the camouflage–bird connection was tenuous: birds in the wild had never been seen eating peppered moths at all. It took another thirty years to move the debate on.

The first person ever to watch a bird snack on a peppered moth was Hazel Kettlewell. It was July 1, 1953, and Hazel was peering through a pair of binoculars at a peppered moth resting on a tree trunk in Cadbury Bird Reserve, a piece of woodland under the smoke of Birmingham. Suddenly, a hedge sparrow leaped up from the bracken, snatched the moth from its trunk, and disappeared from sight again.

It was a momentous occasion. Not only because it was the first time that the crucial observation—that birds *do* forage on peppered moths resting on trees—was made, but also because it happened in the course of one of evolutionary biology's most famous experiments. Hazel was the wife of Bernard Kettlewell, a medical doctor and self-taught zoologist, who had recently been recruited by Oxford University to do experimental work on natural selection and industrial melanism. It was not a random appointment. The energetic, skilled, and knowledgeable Kettlewell had been a long-time friend of the founder of Oxford's informal "School of Ecological Genetics," E.B. ("Henry") Ford. After many years of trying, Ford had finally succeeded in finding enough money to lure Kettlewell from his self-imposed South African exile. He was convinced that if anybody could find the missing pieces of the peppered moth puzzle (do birds really eat the moths? do they really eat fewer of the moths that better match their background? and is the difference really big enough to be the driving force of the wing color evolution?), it was Kettlewell.

So, Kettlewell and his family spent much of 1952 in a trailer in Wytham Woods, a forest that belonged (and still does) to Oxford University. There, they collected and hand-reared some 3,000 peppered moth caterpillars, carefully nurturing them to the point of pupation, and caring all winter for the resting pupae. In June of the next year, shortly before the moment the pupae were expected to hatch, Kettlewell drove up to Cadbury Bird Reserve, with the pupae, carefully packed in gauze, in the back of his Plymouth.

He had chosen Cadbury Bird Reserve because it lay near enough to Birmingham to be covered in thick industrial grime. Setting up a field lab in his caravan, for eleven consecutive days he would be working round the clock, assisted by Hazel. They put individual paint marks on the wings of the moths as they emerged from their pupae, placed the insects onto tree branches, and then at night ran two kinds of moth traps: mercury vapor lights and gauze sleeves with sexually receptive female peppered moths, both tried and tested ways of attracting moths (the latter of course only works for males). The expectation was that, during those eleven days, birds would eat more of the pale, poorly camouflaged moths than inconspicuous black ones, and that this uneven predation would show up as a difference in how many of each category would survive long enough to get into the traps.

It was during one of these days Hazel spotted that hedge sparrow picking off a moth. And more observations followed. Over the next few days they saw several hedge sparrows and robins feeding on both pale and dark *Biston betularia*. They walked their rounds of the forest and recorded which of the moths released in the morning were still on their perch in the evening. For the dark-winged moths, 63 percent were still where they had last seen them, but for the pale-winged moths, this was only 46 percent. The difference was pretty much exactly as large as Haldane had predicted. As Kettlewell wrote in his famous 1955 article "Selection Experiments on Industrial Melanism in the Lepidoptera," this was because "birds act as selective agents, as postulated by evolutionary theory."

Meanwhile, the night-time traps were also reeling in data. Over the eleven days the traps operated, the Kettlewells released 630 (pale as well as dark) male moths, of which they caught 149 in their traps. The recaptures were not equally divided among pale and dark forms. Of the released pale moths, they recovered 13 percent, while more than twice that proportion (namely, 28 percent) of the released dark moths were caught in the traps. Again, it seemed that something was picking off pale moths at a greater rate than dark ones. That something, as now seemed to be confirmed, were birds.

Two years later, Kettlewell performed the counter-experiment, namely releasing some 800 marked moths in an unpolluted forest under the clean Dorset air, and, as expected, found exactly the reverse. Here, it was the dark moths that stood out on the clean, lichen-festooned birch stems, while the pale moths were much harder to spot. Sure enough, the latter seemed to survive better, as they turned up at a greater frequency (14 percent) in his traps than the former (5 percent). This time, he brought another companion to his field camp: Dutch behavioral biologist, and later Nobel Prize winner, Niko Tinbergen, who was gaining fame with his studies of bird behavior, and pioneering the use of cinematography in field biology. While Kettlewell busied himself with his muslin moth bags and mercury vapor lamps, Tinbergen sat in his hide with his camera and shot stunning footage of spotted flycatchers, nuthatches, and yellowhammers feasting on the rich pickings of pale and dark *Biston betularia* that had been laid out for them.

Tinbergen's films and photographs, Kettlewell's papers (he wrote up his Dorset data, as well as a rerun of his Birmingham study, in a second paper in *Heredity* in 1956), and the repeated wheeling out of Kettlewell and his moths by his mentor, Henry Ford, did much to propel the peppered moth as *the* celebrated case study of evolution in progress. By the mid-1960s, *Biston betularia* began to make standard appearances in lectures, documentaries, and articles about evolution, and for the remaining decades of the twentieth century, no biology textbook was complete without photographs of pale and dark moths on ditto tree bark. In fact, this early example of urban evolution is so familiar and worn out that I would not have brought it up here, were it not for a twist in the story that began to unfold in the late 1990s. Chances are that you are vaguely aware of this—that, all the while you were reading the preceding pages, there was this gnawing feeling in the back of your mind that you had heard or read somewhere about something fishy with this paradigm of rapid evolution.

The fishiness started with *Melanism: Evolution in Action*, a 1998 book by Cambridge evolutionary biologist Michael Majerus. The centerpiece of the book is the painting of a much richer picture of the peppered moth case than had ever been done before. Majerus points out some unanswered questions, several of which had been asked by previous authors as well. Do moths always rest on tree trunks or also in places where their wing color does not afford any protection? Since the moths fly by night, couldn't bats, rather than birds, actually be their chief natural enemies? And were the artificially high densities of moths released by Kettlewell in his experimental woodlands really a proper way to study real natural selection? Majerus had intended this critique as a way to urge his colleagues not to lie back and consider the peppered moth case closed, but to take up the gauntlet and start up new, more detailed studies to fill gaps and remove uncertainties. In fact, he himself was working on this.

Rather than stimulating new research on industrial melanism in *Biston betularia*, and much to Majerus's dismay, his book had the unintended effect of casting doubt on the whole story. A book review in *Nature* by geneticist Jerry Coyne had it that "[for] the time being we must discard *Biston* as a well-understood example of natural selection in action," and: "Depressingly, Majerus shows that this classic example is in bad shape." Some of Coyne's colleagues, who knew what Majerus's true intention had been, were surprised by this interpretation of the book. One wrote, "If I hadn't known differently, I would have thought the review was of some other book."

But the damage had been done. Articles began appearing in newspapers bearing headlines like "Scientists Pick Holes in Darwin Moth Theory" and "Goodbye, Peppered Moth." But the worst was still to come. In 2002, journalist Judith Hooper dropped a bombshell titled *Of Moths and Men: Intrigue, Tragedy, and the Peppered Moth*, a well-written and well-researched history of the peppered moth case, in which she dissected the complicated relationships among England's moth men, and insinuated that Kettlewell's experiments had been soiled by his subservient relationship with the Oxford intellectual giants for whom he worked. Basically, she accused him of fraud, aimed at pleasing his superiors. Although she offered no concrete evidence of any wrongdoing, she managed to mount a hatchet job by word choice and guilt by association, and her book was immediately picked up by the creationist community in the US, probably her main intended market. The Institute of Creation Research wrote, "What a wonderful time to be a creationist, when even the supposed best proof of evolution in action is so flimsy that it cannot stand the test of truth."

Hooper's book, together with the brouhaha surrounding his own book, at least served the purpose of spurning Majerus into action. He set out to do a series of massive experiments of the kind Kettlewell had done, but avoiding the pitfalls, and settle the issue once and for all. Place of action: Majerus's own, two-and-a-half-acre garden near Cambridge. When: the years 2002 to 2007. The key players: 4,864 peppered moths (nearly ten times as many as Kettlewell had used for any of his experiments), all resident, of their own accord, in Majerus's garden. (Kettlewell had always mass-reared the moths and then transported them to field sites hundreds of miles away—a procedure that had been criticized, because it flooded the field site with more moths than would naturally occur there, and ones possibly not suited to that particular site at that.)

Another thing Majerus did differently was that, rather than placing the moths on a tree, he allowed them to find their preferred resting site by themselves. Each moth was given a mark and then released, at night, into a large cage wrapped around the trunk and branches of a tree. In the morning, before dawn, Majerus removed the cage, noted where the moth was sitting and checked four hours later whether the moth was still there. If it had gone, he assumed that a robin, hedge sparrow, blackbird, or any of the other insect-eating bird species present in the garden had gobbled it up—something which Majerus, scanning the trees from his garden shed through his field glasses, actually saw happening on no fewer than 276 occasions.

Imagine, just for a minute, the dedication! Majerus possessed twelve cages, so each night he could not release more than twelve moths. That means that, over the six years his experiment ran, he spent more than 400 nights rigging the cages, taking notes, setting his alarm to go off before the early mid-summer dawn, removing the cages, sitting with a cup of coffee and his binoculars behind the window and watching out for birds to swoop in. Let's remember that this work would probably have come on top of his regular teaching and administrative duties at the university—a persistent Herculean effort intended to prove beyond any doubt that the dark-winged peppered moths evolved by natural selection, applied by the conspired action of pollution and predatory birds.

And prove it he did. Or rather, he proved that the peppered moth was, by now, evolving back to its original state. After legislation was put in place in the 1950s and 1960s to curb air pollution, the blackened trees of industrial England had slowly become a thing of the past. The air became cleaner, lichens returned, and the tables were turned on the dark-winged peppered moths. Over much of England, their dusky wings no longer afforded any protection—the reverse, in fact—and bit by bit their advantage waned. As a result, between 1965 and 2005, at more or less the same rate as they had increased a century before, their proportions declined. Today, the dark form is as rare as it was in 1848.

So Majerus's experiments fell in the tail end of this evolutionary decline. In fact, over the six years that he conducted his studies, the proportions of dark-winged moths in his garden dropped from 10 percent in 2001 to 1 percent in 2007. And the outcome of his experiments matched this: each day, around 30 percent of the dark moths were snatched up by birds against 20 percent of the pale ones.

Majerus presented the results of his six-year experiment at a conference in Sweden in 2007, but, sadly, he was not given the time to publish them in the scientific literature. At the end of 2008, he was struck down by a particularly aggressive mesothelioma, to which he succumbed, aged only 54, in January 2009. Soon after he died, his family permitted four of his friends to work up his notes and the slides he had used for his lecture in Sweden and publish them as a paper in *Biology Letters*, which appeared in 2012 under the title "Selective bird predation on the peppered moth: the last experiment of Michael Majerus." The final sentence of the article reads: "The new data, coupled with the weight of previously existing data convincingly show that industrial melanism in the peppered moth is still one of the clearest and most easily understood examples of Darwinian evolution in action."

And so it really is. As a final note of triumph, in 2016 the peppered moth saga was crowned with a paper in *Nature* by a large team of geneticists led by Ilik Saccheri of the University of Liverpool. They showed that the mutation responsible for the black wings was actually a 22,000-letters-long runt of "jumping" DNA that had cut-and-pasted itself into the *cortex* gene that controls wing coloration in butterflies and moths. Detailed dissection of the gene's structure and the adjoining parts of the chromosome showed that industrial melanism in the peppered moth dated back to a single DNA-jumping event that, the researchers calculated, must have taken place in the north of England around the year 1819, "which is kind of smack in the middle of the early part of the industrial revolution," says Saccheri.

The accumulated new evidence of these past few years appears to have vindicated Bernard Kettlewell and reinstated the case of the peppered moth as a genuine textbook example of evolution by natural selection. But in addition, it is also the first recorded case of urban evolution, or more generally, of what has been called HIREC: Human-Induced Rapid Evolutionary Change. It has shown that humans, especially dense urban conglomerations of humans, have the power to exert, on wild animals and plants, novel and unusually strong selection pressures of the order of 10 percent or more. The steep evolutionary rise and fall of the dark-wing mutation in the *cortex* gene of *Biston betularia* is nothing less than a writing on the wall that heralds the coming systemic change of urban nature. With the peppered moth, it was a reversible evolutionary see-saw: a single gene that rose and then fell and that has become a celebrated case because of its simplicity, clarity and, despite the controversy, easy interpretation.

9
SO IT REALLY IS

DESPITE SOME RECENT RUMBLINGS, the peppered moth has regained its rightful place as a textbook example of urban evolution in action—by turning from pale to black during the height of the industrial revolution's air pollution, and back again from black to pale when the worst was over. It has shown that a single change in an organism's DNA, propelled by strong, human-induced natural selection, can kickstart an explosive evolutionary changeover. In this chapter, I will showcase a few more animals and plants that have recently evolved rapid and dramatic changes in their appearance because of adaptation to the urban environment. But first, there is just one more thing I need to point out about industrial melanism in the peppered moth: namely, that it is an urban version of an otherwise common and natural kind of evolution in moths.

In fact, Bernard Kettlewell wrote a whole book about melanic moths. In the UK alone, dozens of species have evolved two or more shades of gray. Not because of soot-covered tree bark, but because each is better suited to different habitats or different parts of the country. In his letter to Darwin, Albert Farn had already alluded to this in the Annulet: "almost black on the New Forest peat; gray on limestone; almost white on the chalk near Lewes; and brown on clay, and on the red soil of Herefordshire." Affording the best camouflage against a particular soil type, natural selection would have pushed for different wing-color genes in each area. Where two areas abutted, the habit of the moths to fly around and mate some distance from where they were born would have spread some pale-wing genes into the New Forest, or some black-wing genes into Herefordshire, but not enough to muddle the geographical color pattern, since the inappropriately decked out moths would always be easy pickings for the local avifauna.

Kettlewell also devoted himself to such "natural" melanism (which he called rural, rather than industrial, melanism). One of his former students, Stephen Sutton, recalls how he joined Kettlewell on a 1960 expedition to the Shetland Isles, where they investigated rural melanism in a moth named the Autumnal Rustic (*Eugnorisma glareosa*): "I was stationed on sand dunes (white background) and other assistants were at intervals up the length of the Shetlands to the blasted heaths of Unst. [The moths were] very pale on my dunes, and dead dark on the open peat of Unst. In the simmer dim of June, gulls would forage through the night and matching your background was a key survival factor."

So, in a sense, the ink blotch of melanic peppered moths that oozed onto England's map in the late nineteenth, early twentieth

century, was no different than the age-old areas with black Annulets on the dark soils of the New Forest, or the "dead dark"
Autumnal Rustics on the Shetland heathland. All had probably
evolved due to mutations that changed wing coloration, and birds
that hunted by eye—although the industrial melanics in the peppered moth had probably evolved much faster, because the landscape changed so quickly and drastically. So fast, in fact, that
we saw it happen before our eyes.

In the peppered moth system, urban pollution caused the moths
to evolve, in essentially the same way—only faster—that moths
had been evolving under natural conditions for millennia, with
birds acting as evolutionary middle-men. But birds themselves
may also be affected by urban evolution. To see that happening, we
should dip into Shakespeare.

In *Henry IV Part 1*, Hotspur is planning to drive King Henry
crazy by letting a starling endlessly repeat the name of Hotspur's
brother-in-law Mortimer: "Nay, I'll have a starling shall be taught
to speak nothing but 'Mortimer,' and give it him, to keep his
anger still in motion," muses Hotspur. In 1877, this obscure
Shakespearean reference to *Sturnus vulgaris*, the European starling, landed the bird a place on the list of animals and plants that
were to join the human colonizers in the US. For in that year,
drug manufacturer Eugene Schieffelin became president of the
American Acclimatization Society, a group of idealists who saw
it as their calling to "improve" North America by releasing "such
foreign varieties of the animal and vegetable kingdom as may be
useful or interesting." And for some unfathomable reason, Schieffelin's particular brand of acclimatization included the bringing
into the USA of every bird ever mentioned in Shakespeare's
works.

Schieffelin's greatest successes were achieved with Hotspur's
starlings. In 1890 and 1891, he had some eighty breeding pairs

shipped from England and released them in New York's Central Park. Instead of sitting around repeating royal names, the birds wasted no time and immediately proliferated into the vacant niche of winged inhabitant of American cities and villages. Researchers have calculated that, from their point of release, they multiplied and spread with a speed of some 50 miles per year, hopping from town to village to hamlet. In 1920, they occupied the entire US east coast. By the end of the Second World War, they had crossed the great plains. In the 1960s, they had established themselves on the west coast, pushing on into the interior of Alaska by 1978. Today, there are about as many starlings as there are people in North America.

Clearly, bolstered by its Shakespearean mandate, *Sturnus vulgaris* decided to be, and not not to be. But to establish itself in all those burgeoning American cities may have placed demands on the starling's nimble body. And those demands, two Canadian researchers reckoned, might be different from what had shaped the bodies of the original English colonist-starlings. To check this, they consulted the bird collections of eight natural history museums in North America and took measurements on the shape of the wings of 312 starlings, stuffed in the 120 years since their departure from Central Park in 1890.

The scientists, Pierre-Paul Bitton and Brendan Graham from the University of Windsor in Canada, discovered something interesting. Over time, they found, the starlings' wings had gradually become more rounded, because the secondary flight feathers (the ones at the bird's "lower arm," closest to the body) had become elongated by some 4 percent.

Now, the shape of a bird's wing is not something that evolution can mess with with impunity. It is very closely wedded to a bird's way of life. Long pointed wings are better for fast flying in a straight line, while short, rounded wings are good for making rapid turns or for quickly taking off. That's why the dive-bombing peregrine falcon has the former, but an aerial acrobat like the lapwing has the latter. It is precisely this quick-response benefit of more rounded wings that may be one of the reasons that the settler starlings evolved. In those 120 years, the human population in western North America (the part of the continent that the starling expanded into) grew almost fifty-fold. What were tiny settlements when the starling arrived, blossomed into metropolises in a matter of decades. And with urbanization came new dangers for urban birds: cats and cars. It is quite likely that this is what caused the American starlings to evolve a wing shape that helped them get out of the way of a pouncing cat or a speeding motorcar hurtling toward them.

In the case of the starling's rapid wing evolution, we can only speculate about what exactly caused it. But in the evolution of roadside American cliff swallows, we know for sure.

Blessed is the bird to whom a biologist devotes his or her entire life. In the case of the American cliff swallow (*Petrochelidon pyrrhonota*), it is even his *and* her. Since 1982, Mary Bomberger-Brown and Charles Brown have spent every spring studying colonies of these birds in Nebraska. Around the time they began their work, the swallows, which normally build their gourd-shaped clay nests on crumbly rocky overhangs and sandy cliffs, had just taken up the habit of colonizing firm, newly built concrete highway bridges and road-side culverts. "We built them a better cliff," says Bomberger-Brown. Over the years, some colonies grew as large as a staggering 6,000 nests, all suspended from these artificial structures. And each year, the two biologists monitored the colonies, driving along the same roads for the same number of field days, using mist nests to catch the swallows, measure them and adorn their legs with tiny numbered rings. They also made it a habit to pick up any dead cliff swallow they found along the roadside, and to take its measurements—such as wing length.

As so often in scientific research, their meticulousness, stamina and utter immunity to boredom paid off in the end. In a two-page article in *Current Biology* in 2013, they assembled all the data that their thirty years of calipering swallow wings had yielded. In the 1980s, when the birds had just begun nesting on the roadside structures, all birds, dead or alive, had wings that were about the same length: around 10.8cm. But as time went on, they found that the living birds' wings grew shorter, by about 2mm per decade. Not much, and perhaps not really worth noticing if their measurements on the roadkill had not shown the exact opposite pattern: by the 2010s, the wings of dead birds by the roadside were about half a centimeter longer than those of live birds still happily flapping along. Also, even though the pressure of traffic had remained the same or even increased, the numbers of dead birds declined by almost 90 percent.

The conclusion was inescapable: only cliff swallows with wings short enough to take off vertically from the tarmac to escape an oncoming car had managed to get away and spread their short-wing genes in the gene pool. The tardier long-winged ones ended up as ex-swallows on the hard shoulder, their long-wing genes excluded from the gene pool. And, as the surviving swallows became ever better adapted at evading approaching vehicles, the number of casualties plummeted.

On the other side of the Atlantic, in southern France, unforgiving tarmac also drives evolution. Not birds, in this case, but plants. Pierre-Olivier Cheptou, a botanist at the Montpellier's branch of the National Center for Scientific Research (CNRS), has been studying weeds on the city's pavements. Or rather, in the one-square-yard quadrats of soil around trees planted along those sidewalks. As Cheptou writes in one of his papers, "these patches are widely distributed (several thousand in the city), and regularly spaced, from 5 to 10 m[eters] from each other depending on the street." Indeed, as I descend into Montpellier on spaceship Google Earth, the square snippets of soil become visible on my computer screen all over town. Like tiny bits of Versailles, the geometric pattern along Rue Auguste Broussonet, Avenue Henri Marès, and Chemin des Barques looks like a city-sized ecological experiment. And that is exactly how Cheptou and his colleagues used it.

Almost a hundred different species of wild plants grow in these patches of soil. One of them is the hawksbeard (*Crepis sancta*)—it looks a bit like a dandelion, but with the yellow flowers sitting on top of multiple branched stems, rather than a single one. Like dandelion, after blossoming, the flower changes into a head of fluffy seeds. Most of these seeds are small and light and have a delicate parachute-like umbrella attached to them. Others, however, lack the parachute and are much heavier. This seed-duality is the hawksbeard's way of having your cake and eating it, seed-dispersal-wise. The heavy seeds fall straight to the floor, where they are sure to find a suitable bit of fertile soil at the parent plant's root. The parachute-seeds, on the other hand, will be lifted into the air by a gust of wind or a child blowing them off their pedestal, and float on the wind until they drop somewhere far away from their parent. With a bit of luck they find a suitable, vacant bit of soil.

At least, that's how it works with hawksbeards growing in the wild. In *Centre Ville* Montpellier, however, "a bit of luck" is hard to come by. Barring a few cracks in the pavement, those one-by-one-yard squares of soil strewn with dog poo, discarded candy wrappers, and Gauloises butts, are the sole spots suitable for seed germination. So, even though the plant owes its colonization of the city center to stray parachute seeds that were lucky enough to land on soil, once it had colonized the city's streets, the heavier seeds that fall straight down were the hawksbeard's only sure chance at reproduction.

For city hawksbeards, therefore, it would make evolutionary sense to produce more heavy seeds and fewer parachute seeds. And that is precisely what they do, Cheptou found. He sampled hawksbeard seeds from patches of soil along the sidewalks of seven city center streets. Then, he did the same for hawksbeards growing in four meadows and vineyards in the countryside around the city. He brought all of the seeds to his lab at CNRS and grew them in the same greenhouse, under the same conditions.

When the plants finished blossoming, he counted the numbers of heavy seeds and parachute-seeds in all of the flowerheads. Plants grown from seeds collected in the city, he found, produced almost one and a half times as many heavy seeds as countryside plants. And for the numbers of parachute seeds the reverse was true. In other words, in the city, the plants had evolved to sacrifice the parachute-seed production in favor of the heavier seeds. Based on the loss of progeny and the degree to which seed production is determined by genetics, Cheptou was able to calculate that it must have taken about twelve generations for this evolutionary change to take place. The plants have one generation per year, and since the streets where Cheptou plucked them had been re-paved between ten and thirty-three years previously, this seems to be an example of extremely rapid, urban evolution.

Again, as rapid and inner-city urban as it may be, what evolution does here is nothing unprecedented. Basically, those Montpellier hawksbeards have become island plants. The one-square-yard patches of soil form an archipelago within a sea of impervious tarmac and concrete, and the hawksbeard's investment in seed types evolves the same as it does in plants on real islands in the ocean. In fact, in the 1980s, Martin Cody and Jacob Overton of the University of California in Los Angeles found something very similar in the seeds of another dandelion-like weed, *Hypochaeris radicata*, or cat's-ear, on twenty-nine tiny islands in the Barkley Sound of Canada. Unlike the hawksbeard, these plants only produce a single type of seed, suspended below a parachute. But when Cody and Overton measured the sizes of the seed and those of their parachutes, they found that the island plants produced much heavier seeds with tinier parachutes than plants of the same species growing on the Canadian mainland. The explanation is the same as for Montpellier's hawksbeards: all the plants that produced light seeds saw their offspring fly off into the sea, and were punished by natural selection, which favored the evolution of plants producing ever heftier seeds suspended from ever punier parachutes!

Anolis lizards, too, have recently added their own urban twist to their already formidable conventional evolutionary portfolio. Dwarfed only by the 2,000 cichlid fish species of the great African lakes, the anoles of the Caribbean, Central and South America form one of the greatest evolutionary diversifications of the vertebrate animals. Literally a textbook case of "adaptive radiation," they have fanned out into some 400 species, each with its own specialization and locale. They can be tiny, just a few centimeters long, or giant, up to almost one and a half feet. They can be pretty green or turquoise, gray or brown patterned, with a snout that may be snub-nosed like *Anolis nitens*, or long enough to warrant the nickname "pinocchio lizard" (*Anolis proboscis*). There are chameleon-like stocky ones that crush snails and moth pupae. There are sleek water-diving ones that catch crayfish. And then there are all manner of tree-dwellers: the tree trunk anoles have long legs to jump on the ground and chase prey. The twig dwellers have short legs to hold onto their twig. The canopy dwellers have large toe pads to stay stable on the slippery canopy leaves. (Besides geckos, anoles are the only lizards that can hang by a single toe.) There are also grass species with extremely long tails and lengthwise stripes—they look almost like a blade of grass when you see them. All of these types of anole have evolved repeatedly on different islands. "They've done it in quadruplicate across the four islands of the Greater Antilles!" exclaims Jonathan Losos of Harvard University, the world's foremost *Anolis*-evolution expert.

Although it has taken 50 million years for this diversity to come about, this does not mean that anoles evolve slowly. In a famous experiment started in 1977, researchers caught *Anolis sagrei* on the small island of Staniel Cay in the Bahamas, measured all of them and then released handfuls of males and females on fourteen small, previously anole-free islets in the vicinity. After about ten years, they went back and caught and measured the colonists' descendants. They discovered that, compared to the original Staniel Cay population, the settlers had evolved. Overall, their legs had got shorter and their toe pads bigger—the more so if the vegetation in their new home was more grassy. This is because to run fast on the big tree trunks of their original island, Staniel Cay, the lizards needed long legs and narrow toe pads. However, to get about on the thin twigs and blades of grass of their new homes, it's important to cling on and the best way to do that on those narrow slippery stems is with short legs and sticky toes. (All this was duly confirmed in the lab where lizards were chased up inclined racetracks of various widths.)

The Staniel Cay experiments showed that *Anolis* lizards are capable of fast evolution. How fast? Evolutionary biologists have a unit for that: the darwin—one darwin being roughly an increase or decrease of 0.1 percent per 1,000 years. The Staniel Cay lizards evolved with a rate of between 90 and 1,200 darwins (that's 1.2 kilodarwins, if you like that mental image). Not a world record, but quite impressive, as evolution goes.

This opened up perspectives for urban evolution, too. And *Anolis* does not eschew the city, as another *Anolis* researcher, Kristin Winchell, found out. She decided to work on the Puerto Rican Crested Anole (*Anolis cristatellus*), a small anole of the tree trunk type that occurs all over the island of Puerto Rico, in the countryside as well as in the city. Winchell picked residential neighborhoods in three of the island's biggest cities and for comparison chose a forest on the edge of each of these cities. At each of these six sites, she caught fifty lizards with a silk noose, a kind of miniature lasso on the end of an extendable fishing rod, then snipped off a small piece of the tail, stuck the animals in a portable x-ray machine, scanned their feet on a flatbed scanner, and wrote a little number on their scales. After that, the perplexed lizards were returned to their original perch (and are probably still telling their grandchildren about that day they were abducted by aliens).

Winchell's project (published in *Evolution* in 2016) showed a clear sign of urban evolution: the urban lizards always had longer limbs and more lamellae on the underside of their toe pads. And yet, the DNA from their tail tips showed that the urban lizards of each city were most closely related to the local forest lizards, not to those from the other cities. So, the differences had evolved three times independently. To check that the limb and toe pad differences were really genetic, Winchell also took a hundred urban and forest lizards to her lab in Boston, waited for them to lay eggs, and let all the offspring grow up under the exact same conditions. In these expat lizards, too, the ones from urban parents had longer legs and more toe-pad lamellae than those born from forest parents, which proved that the difference was truly in their DNA, and not simply induced by their growing up in a city environment.

All in all, this showed that the urban lizards had adapted to the urban kinds of perches they use: usually walls where you need to run quicker and farther to get out of harm's way than on a tree trunk. Also, the urban perches were often very smooth (painted concrete walls and metal), which required heavier-duty toe pads to remain clung. In fact, a decade previously, other researchers had already discovered that urban lizards fall down more often (and usually onto a much harder surface) than forest lizards, leading to injury or even death. For them too, the old adage is true: most accidents happen in and around the home.

The *Anolis* lizards, hawksbeard plants, starlings, and swallows show us that urban evolution is fast, observable, and actually quite straightforward. In the following chapters, we're going to see that urban evolution can also be complex, tortuous, and counterintuitive. But first we must pay some attention to the matter of fragmentation. How can evolution ever gain a foothold if the urban gene pools are cut into miniature shreds by all the barriers we create in our cities?

10
TOWN MOUSE,
COUNTRY MOUSE

L ONG, THIN NEEDLES STICK UP FROM
the marble heads of Reine Mathilde, Marie Irè d'Écosse,
and all the other statues around the Jardin de Luxembourg in
central Paris. Though this gives a fleeting impression of them all
being connected by wireless to the great mother ship of French
aristocracy in the sky, the "antennae" are in fact meant to keep the
park birds from landing and covering their regal heads in unsightly
droppings. Still, one ring-necked parakeet is undeterred and briefly
perches on the crown of Reine Berthe, demonstratively deposits a

dropping on her cheek, and then flaps off to join a loudly screeching group of parakeets zigzagging among the tall plane trees.

Parakeets in Paris . . . It could be the name of a hypnotic Matisse
painting, but since the 1970s it has been a very realistic image for
the French capital. In fact, the ring-necked parakeet (*Psittacula
krameri*) is one of the birds that has been most successful in invading cities in Europe (on a smaller scale, also in Japan, North America, the Middle East, and Australia). Originally hailing from India
and Africa, the bright green parakeet with its red beak and long
tail (and, in males, a black-and-pink cravat and an azure tail) has
been hugely popular in the caged-bird trade for much of the twentieth century. The numbers in trade have been so large (almost
400,000 imported into western Europe since the 1980s) that on
the one hand the original populations in the tropics have declined,
but, on the other hand, the species has established itself in cities all
over Europe as booming populations founded by the inevitable es-
capees.

The birds have also been granted pardon on a number of nota-
ble occasions. Jimi Hendrix reputedly set free a pair on Carnaby
Street in London in the late 1960s, which some say helped create
the now huge London parakeet population. And by releasing forty
parakeets in 1974 "because Brussels could use some more color,"
the owner of a Belgian zoo single-handedly founded the country's
entire population, now numbering some 30,000 birds—more than
4,000 of which sleep every night on the appropriately named De
Neck Street in Brussels. This may have been a bit more green, red,
orange, yellow, and blue than zoo director Guy Florizoone (note
the name) had intended.

Like the house crow, the ring-necked parakeet is another example of a tropical bird that, seemingly effortlessly, has set itself
up as an enterprising urban bird in northern Europe. The birds
benefit from the urban heat islands and the fact that there is food

to be had in cities in winter (they particularly like to monopolize the strings of peanuts that people put out for smaller songbirds). It also helps that the species' native range includes the foothills of the Himalayas, so it may already have been pre-adapted to cold spells. Thanks to all these factors, the squawking flocks of fast-flying parakeets are now a familiar sight to most European urbanites. In Paris, I see them squabbling over access to tree hollows in Versailles, and zipping in noisy groups across the evening sky high above Boulevard Montparnasse, on their way to their favorite communal sleeping trees in Parc Montsouris. They fly around in the Jardin des Plantes and among the tall plane trees of Jardin de Luxembourg. Ariane le Gros, a biologist with the Muséum National d'Histoire Naturelle, says: "You find them in more and more parks of Paris, as the population expands really fast."

Le Gros is one of the biologists who have begun studying how the gene pools of urban species are structured, using an approach known as "phylogeography." Phylogeography began in the 1980s as a way to tell the evolutionary history of natural populations of animals and plants. It usually involves looking at a large number of "markers," variable bits in the DNA of a species, for a lot of specimens from different parts of a species' area of distribution. Phylogeographers can then use such rich information on the genetic make-up of a species to trace back its history. They can calculate what has been the mostly likely colonization route. Or they can see if a population has perhaps been reduced in size at some point in the past, and, if so, how long ago. For example, even if nobody had ever found a single human fossil, phylogeographic analysis of DNA of people living today would still tell us that we evolved in Africa, the routes by which we colonized the rest of the planet, which mountains and deserts blocked our progress, and roughly how many colonists had been involved at each migratory step. It would tell us how long ago each of these steps was taken. It would also tell us, for example, that the gene pool of northern Europe is well mixed, thanks to its long history of travel and trade and intermarriage between people from different parts of the continent. The interior of New Guinea, on the other hand, would show a gene pool much more fragmented by the impassability of the mountainous, thickly forested country. Phylogeography, in other words, is a way to peek into a species' past via its present-day genes.

In recent years, phylogeographers have begun plying their trade on urban species. By looking at the DNA of urban flora and fauna, they are able to answer many crucial questions that cannot be answered in any other way. What Ariane Le Gros wanted to know was, for example, where do the parakeets of Paris come from? And also: do they form a single well-mixed population or are they split into separate tribes? To get answers to those questions, she caught about 100 parakeets in private gardens close to the Parc de Sausset in the north of Paris and near Parc de Sceaux in the south of the city. Le Gros took DNA samples from blood and the roots of plucked breast feathers and used these to get eighteen "markers" on the birds' chromosomes and carry out a phylogeographic study.

To her surprise, her analysis told her that the south-Paris and north-Paris parakeets are as different from one another as they are from those in other European cities, which leads her to conclude that the Paris parakeets come from at least two different stocks. This means that parakeets were released or escaped at least twice in Paris or—less likely, thinks Le Gros—that one or both of them have flown in from elsewhere (the parakeets from northern Paris, for example, are quite similar to those that inhabit Marseille, whereas those from southern Paris are genetically unique within Europe). Either way, what is clear is that they did not mix much since they settled in their respective quarters of the city. This may seem surprising, given that they are fast-flying birds that, it appears, could easily traverse the twelve miles in between.

Not quite, says Le Gros. Even though the birds could fly up to nine miles in a day, they don't really like to cross built-up areas, because, unlike many other urban birds, they cannot live without trees. They need them for roosting at night, since their feet are not made for perching on rocky ledges like pigeons. And they need them for their nests, because, even in the city, they insist on breeding in tree hollows. In a thoroughly stony city like Paris, where even famous parks like Place des Vosges or the Tuileries consist of little more than swathes of sandy path with a few rows of dusty trees, and hardly qualify as green spaces, suitable habitat is hard to come by. The many miles of treeless cityscape that separate the parks of Paris where the parakeets *do* feel at home, are probably sufficient to keep the city's populations from genetically blending.

And that's just birds. Think of more ground-bound city animals, and the phenomenon of fragmentation gets even more acute. Bobcats (*Lynx rufus*), for example. These smallest of the world's four species of lynx occur throughout most of North America. They are about twice the size of a domestic cat, and although they were, and still are, hunted intensively for sport and for their soft and attractively patterned fur, they have survived quite successfully and in recent years have been making a bit of a comeback. More and more bobcats are beginning to set themselves up in suburbia, where they sometimes get chased up trees by domestic dogs. They even dare to penetrate the inner city now and then, although their favorite habitat is forest edge with plenty of rabbits and rodents.

In Southern California, wildlife biologist Laurel Serieys did a phylogeographic study of the bobcats within Los Angeles and

the region north and northwest of it. This vast area is a patchwork of city sprawl, agriculture, vegetated hills, and residential
estates. And roads. Lots and lots of roads. The 56-by-31-mile
area that Serieys studied is cut into four quarters by two of the
US's busiest freeways: Route 101, running east–west, and Inter-
state 405, running north–south. Between them, these ten-lane
highways, the scenes of many a Hollywood car chase, support
some 700,000 vehicles per day, throwing off secondary and ter-
tiary roads all over the place. But while they connect the human
populations, these arteries of traffic, Serieys found, are very effi-
cient at disconnecting the bobcats from one another.

Using an arsenal of animal-friendly capturing devices (padded
foothold traps, cage traps, and box traps), supplemented with occasionally encountered roadkill, she and her colleagues were able
to obtain DNA samples from nearly 400 bobcats prowling the
area. The cats' genetics clearly betrayed how the roads carve up
their habitat. Bobcats in the section east of the I-405 and south of
Route 101, on the northern edge of Los Angeles (Beverly Hills,
Hollywood, and around the Hollywood Bowl) formed one tribe,
genetically very different from the ones north of Route 101, living
in the planned community of Thousand Oaks, until 1969 home of
Jungleland, where they filmed *Tarzan*. The Thousand Oaks bobcats again have different DNA from those south of Route 101, but
west of the I-405, in the non-urban Santa Monica Mountains.

While bobcats cannot cross freeways, they easily traverse
smaller roads. Hence, the genetic structure of their population
is determined only by really big highways. But tinier mammals,
like mice, are impeded even by lesser streets. In New York, zoologist Jason Munshi-South of Fordham University has made a
name for himself by mapping the exact phylogeographic structure of the city's wild mice.

When I first met Munshi-South, back in 2005, he was a lean

tropical forest ecologist doing his doctorate on small jungle mammals from the same university in Malaysian Borneo where I was working. Now, when I meet him again over Skype in 2017, I see a transformed man. He has exchanged his jungle gear for a shirt and pullover, his tropical sleekness for urban ring-bearded solidity, and chats with me from behind a reliable oak desk in his Fordham office. But the rodentology paraphernalia on the wall behind him betray that he is still very much a field zoologist. Except that the "field" is now no longer the tropical rainforest, but New York City's parks.

"It was originally a side project," he says of his first forays into studying the white-footed mouse (*Peromyscus leucopus*), back in 2007. "I met some people at a meeting who gave a presentation on citizen science and small mammals in New York City, and that's how I became interested. So I rounded up some undergraduates and started trapping that first summer."

The white-footed mouse (big beady black eyes, a grayish-brown coat with strikingly white belly and feet), Munshi-South reminds us, "is not the mouse you find running around your apartment. This is a native species, been here long before humans." A few hundred years ago, in the time that Eric Sanderson re-creates in the Mannahatta Project of Chapter 1, the whole New York area would have been covered in forest and meadows, with white-footed mice everywhere. They would have formed one continuous population, a well-mixed gene pool with DNA flowing freely through it, as is still the case today in non-urban areas on North America's east coast. In early twenty-first-century New York City, what remains of the mice and their original habitat are populations marooned in isolated patches that have stayed green and we now call parks. Central Park in Manhattan and Prospect Park in Brooklyn are the largest, but the native mice also live in small ones like Willow Lake in Queens.

Although they are imprisoned in these parks (they will only travel under cover of vegetation and most parks are not connected by anything that's green), the mice are doing quite well there, especially in the smallest parks, which are too small to support predators like owls or foxes. "There also aren't as many competitors," says Munshi-South. "Particularly things like deer . . . They really decimate the understory in a lot of places where they've become abundant, and that reduces the resource base for white-footed mice considerably. But in the city, I mean, they don't really have many competitors."

As a result, the New York City parks are well stocked with white-footed mice, and have been so since these parks became isolated by urban development around the end of the nineteenth century. As Munshi-South and his students found after they trapped hundreds of mice in birdseed-baited cage traps in fourteen of the city's parks, those 120 years or so have been enough for the mice in each park to evolve their own park-specific DNA. For every mouse they caught, they snipped off the last one centimeter of the tail before releasing the animal. This does not harm the mouse much, and it gave the researchers enough tissue to run their genetic tests. Those tests showed that the mouse population in virtually each park, even ones right next to each other, had its own genetic signature. This is something that in the wild is normally only found across much larger areas, like entire states. "Someone could give us a mouse, not tell us where it was from, and we could determine what park it came from. That's how different they've become," he says.

What the white-footed mice in New York City parks show, just like the bobcats around Los Angeles and the parakeets in Paris, is that urban environments are often so patchy that the gene pools of urban wildlife get split into a multitude of tiny slivers. This is not surprising, in a world with 22 million miles of paved road, where for one-fifth of the land surface the road density is so great that no roadless areas of smaller than a half square mile remain. Usually, it's those long, linear barriers like paved roads, but also railways and walkways, the traffic they carry or the buildings along them that bisect these gene pools, keeping animals or plants and their genes from crossing freely.

Sometimes, it's this infrastructure itself that *is* a species' habitat, like in the case of the London Underground mosquitoes of the preface, where each tube line has its own mosquito population. Or the cellar spiders (*Pholcus phalangioides*) that Martin Schäfer of the University of Bonn studied in buildings in five European cities. He discovered that the spiders living in different rooms within one building jointly form a single gene pool, but that each building is a separate gene pool: the spiders move chambers, but rarely move house.

The received wisdom among biologists is that such gene-pool fragmentation is bad for a species' chances of survival. The idea is that in these small, isolated populations, there's a lot of inbreeding: mating is among relatives and if that relative happens to carry a genetic defect, chances are that you carry it too, and so will your joint offspring. Also, genetic variation disappears because of chance events: if a genetic variant is carried by 5 percent of the animals in a population, then in a big population, 5 percent may still mean hundreds of animals. Not likely that those will all expire without leaving offspring. But in a small population of a few dozen animals, the handful that carry this gene might, by some chance misfortune, all fail to breed one year, and take this rare gene with them to their graves. Such a tendency for the genes of a small population to become ever more uniform is called "genetic drift." Drift and inbreeding deteriorate the "genetic health" of a population. Genetic diseases may gain a holdfast, and the loss of variability may mean that the population cannot adapt if conditions change.

This potential problem is why conservationists are always going on about corridors and connecting populations of endangered animals. It is probably why many species eventually cannot sustain themselves in the subdivided environment that cities constitute. But while they are still hanging on, the randomness of drift and inbreeding causes each isolated population to be dealt a different mix of genes. This is how the signal of genetic fragmentation can be picked up by scanning the genomes of bobcats, parakeets, white-footed mice, and any other species whose gene pool is no longer happily sloshing around.

According to Munshi-South, not all species with fragmented gene pools succumb. "There's these other species, especially the ones we don't think about, the ones that are just there . . . that I find really interesting." His white-footed mice are one of these survivors. Despite being so subdivided that each city park has its own genetic signature, the mice seem not to be suffering from the ill effects of inbreeding or genetic drift. If anything, they appear to be thriving. "I think species that can achieve relatively high population densities in a number of places in the city will generally do okay."

The reason that the mouse population in each park has a distinct genetic make-up, says Munshi-South, is probably not only because of inbreeding and drift, but because of what he calls local adaptation. Each park has its own isolated band of white-footed mice. And since the mice are not going anywhere, nothing stands in the way of them evolving to suit the local conditions precisely.

To study this exciting possibility in more detail, Munshi-South and his student Stephen Harris undertook an innovative genetic project. They caught mice in several New York City parks, and also in some rural sites outside the city. The aim of this new study was not to just look at a few random markers in the genome, but to study a large number of actual genes active in the rodents' organs. Sadly, to do so, these mice had to sacrifice more than just a snippet of their tails, for urban science. The researchers killed each captured mouse, removed liver, brains, and gonads, and extracted all the so-called messenger-RNA from these organs. Messenger-RNA (or, "mRNA") is what a gene is copied into before the cell uses its code to produce a protein. So, the pool of mRNA extracted from an organism tells you which genes are actively being used in the body, and what their exact DNA-codes are.

Then, from this huge pool of genetic information, they picked out all the genes that were more different across parks than expected by chance, since those were the ones that apparently had evolved in different directions in different parks. In Central Park, for example, the mice had a distinctly aberrant AKR7 gene. This gene takes care of neutralizing aflatoxin, a toxic and cancer-promoting substance produced by a fungus that often grows on mouldy nuts and seeds. For some reason (perhaps discarded snacks?), the Central Park mice seem to be more exposed to this. Another gene that had evolved in a striking manner in Central Park was FADS1, which plays a role in dealing with high-fat diets—again a tell-tale sign that these white-footed mice had evolved to manage typical Central Park food stuff. Other genes were noticeably different in other parks, and most were either diet-related, or had to do with exposure to pollution. There were also various immune-related genes, which makes sense, says Munshi-South: "Very easy to spread disease when you're in a small population."

Cut to the Hollywood bobcats. There, one chunk of the population subdivided by freeways also seems to have been able to

evolve their immune system. Between 2002 and 2005, the bobcats in the city of Thousand Oaks, cut off from the rest of the
population by Route 101, suffered from an epidemic of mange, a
debilitating skin disease caused by parasitic mites. Research by
the National Parks Service showed that the illness mostly struck
bobcats that had already been weakened by exposure to rat
poisons, which are being used liberally in households and by
exterminators. The poisoned rodents were eaten by the bobcats,
weakening the bobcats' immune systems, which made them sus-
ceptible to mange. This disease can be fatal in the long run, and it
was fatal to so many bobcats that it created a few years of very strong
natural selection, with annual survival rates falling from nearly
80 percent to only around 20 percent. This caused the immune
system to evolve so quickly that the signal could even be picked
up in the genetic data collected by Laurel Serieys. She found that
the bobcats captured before the mange epidemic had a very different set of so-called MHC and TLR genes than after the epidemic. These genes produce proteins that recognize disease-causing
micro-organisms, such as the mites themselves or the bacteria that
enter the skin after it has been breached by the tunneling mites.
Apparently, only bobcats with just the right combination of
immune genes survived the onslaught and made it through the
bottleneck, changing the genetic make-up of this section of bobcat territory for good.

Granted, perhaps the mange epidemic struck bobcats living
north of the 101 because they were already weakened by genetic
drift and inbreeding. But at the same time, the smallness of the
population also allowed it to adapt very rapidly to a challenge
that faced its specific locale. And that is something that may not
have been possible in a larger population, because non-adapted
genes would have come flowing in from all directions. Same with
the white-footed mice: adapting to a specific Central Park envi-

ronment is only possible if Central Park mice are sufficiently isolated from the mice in other parks. Even in Paris the northern parakeets have slightly different head and wing shapes from the southern ones. This could be because the birds that founded them were already different, but it could also mean that subtly different conditions in the different quarters of the city have caused the populations to adapt to the local situation.

For Munshi-South, this change from seeing genetic fragmentation as the bane of urban wildlife to viewing it as an opportunity for each fragment to adapt to the demands of the local neighborhood is a tantalizing paradigm shift. "It's a really interesting question," he says. "Right now, what I want to do is go look at a whole range of populations, several in the city and several along a gradient of suburban and rural, and see whether [local adaptation] is generalizable across all the populations across New York City. And once we have the technology we can look at different cities. I think that's really where urban evolution needs to go. I think that's an open question and a really important one."

11
POISONING PIGEONS IN
THE PARK

DOUGLAS ADAMS, IN *SO LONG, AND
Thanks for All the Fish (the fourth book in his famously
funny Hitchhiker's trilogy) at one point describes how protagonist
Ford Prefect has a dream about New York's East River, so "ex-
travagantly polluted" that new life forms are emerging from it and
demanding welfare and voting rights. I like to think that the
late Adams, also an amateur zoologist and conservationist, would
have been tickled by how reality approaches his science fiction.

While many countries today are cleaning up their act and beginning to outlaw the wanton use and release of pollutants into
their densely populated lands, we have to face the fact that pollu-

tion can never be completely avoided in human habitation. The sheer density and intensity of the smorgasbord of human activities simply means a continued release of substances at higher than background concentrations. And while the most noxious pollutants are replaced by less evil ones, and their use and dissemination is regulated and channeled as much as possible, wild animals and plants in urban environments *will* come across a different, denser, and more varied set of chemical compounds than their brethren in unspoiled places. And all these things can become spanners in the works of the smoothly functioning physiology of animals and plants.

So, coping with a varied and constantly changing landscape of pollutants is one of the many requirements for urban species to survive in the city. Over half a century ago, when Rachel Carson wrote the opening pages of *Silent Spring*, her famous 1962 assault on the pesticides that had become so commonplace so quickly, there was still no reason to be other than despondent about this. She writes:

> Given time—time not in years but in millennia—life adjusts [. . .]. For time is the essential ingredient; but in the modern world there is no time. [. . .] The rapidity of change and the speed with which new situations are created follow the impetuous and heedless pace of man rather than the deliberate pace of nature. [. . .] To adjust to these chemicals would require time on the scale that is nature's; it would require not merely the years of a man's life but the life of generations.

Today, we know that nature, when faced with pollutants, is not always as lame as Carson feared it to be. Potent pollution may cause species to evolve their way out of the noxious quagmire, and many animals and plants have been able to do what New York

City's white-footed mice have done. That is, tinker with their phys-
iological works if pollution throws a spanner in it.

One animal that famously tinkered with its organismal cogs
and wheels in the face of pollution is the mummichog, a fish with
a name that could easily have sprouted from Douglas Adams's
mind—especially given that it was the first fish ever to fly in space.
Known to ichthyologists as *Fundulus heteroclitus*, the mummichog
is a sturdy brackish-water fish about the size of your index finger,
with pretty silvery speckles on an olive-brown background. It lives
along the North American east coast, where it can be found in
estuaries and marshes from Florida all the way up to Nova Scotia.
This vast range already betrays its tolerance and hardiness, which
is one of the reasons why it has been a favorite subject in all manner of experiments since the late nineteenth century. In 1973, it
was even chosen as a passenger on Skylab for experiments on balance and orientation in zero gravity.

Nature has played its own experiments on the mummichog.
Since its area of distribution also comprises some of North America's biggest cities and busiest ports, this little fish has seen its fair
share of polluted environments. It wallows in the muddy bottom
of New Bedford Harbor in Massachusetts, and the port of Con-
necticut's largest city, Bridgeport: filthy industrial mudholes that
contain up to 20 milligrams of PCBs (polychlorinated biphenyls)
per kilo of sediment—the result of decades of twentieth-century,
unfettered release of industrial waste straight into the water.
Twenty milligrams may not sound like much, but PCBs, once used
ad libitum in cooling, lubrication, printing, and a whole range of
other applications, are among the nastiest and most persistent compounds that the previous century has produced. Other compounds
with melodious names but similarly malign outcomes that the
fish's urban habitat are rich in are polycyclic aromatic hydrocar-
bons (PAHs).

The reason that PCBs and PAHs are so troublesome is that they latch on to a type of protein called AHR, which stands for aryl hydrocarbon receptor. These proteins, in humans as well as in fish, act like switches that turn on and off programs of embryo development. When an animal is faced with high levels of PCBs and PAHs, these molecules are constantly tampering with AHR, so that programs are switched on too early or fail to switch off in time. This results in birth defects, especially problems in the development of the heart and blood vessels. Baby mummichogs exposed to PCBs often get hemorrhages in the tail, inflated or underdeveloped hearts, and usually die in mid-development. In fact, mummichogs are among the fish species that are the most sensitive to PCB and PAH pollution.

That is, *average* baby mummichogs. But the mummichogs puddling about in the poison-laden mires of Bridgeport, New Bedford Harbor (and at least two more heavily polluted portside cities on the North American east coast) aren't average. They've evolved a way to cope with chemical foulness.

Andrew Whitehead, a biologist with the University of California in Davis, has been studying the fish's evolutionary agility by comparing their properties in heavily polluted sites (so-called "Superfund" sites, earmarked for clean-up under the US federal program of that name) with those in nearby, pristine waters. Some 40 miles to the southwest of New Bedford Harbor, for example, lies the pleasantly unspoiled Block Island, with PCB levels almost below detection level, a full 8,000 times lower than in New Bedford. And 9 miles south of the toxic cocktails of urban Bridgeport, on the other side of Long Island Sound, mummichogs are happily flapping about in the beautiful lagoon of Flax Pond, devoid even of trace amounts of PCBs.

At these and two more pairs of polluted-versus-unpolluted places, Whitehead caught mummichogs and, using a general

DNA test, checked if the fish from each pair were each other's closest relatives. This panned out quite nicely: the fish from Flax Pond and nearby Bridgeport shared a common ancestor as did those from adjacent New Bedford and Block Island. But that's where the similarities ended, because in many other respects, the mummichogs from each polluted site had evolved drastically away from their relatives in the nearby pristine environments. First of all, Whitehead showed in lab experiments that they were resistant to normally deadly levels of PCBs. At PCB concentrations that would have made the Block Island fish go belly-up ten times over, the hardy New Bedford ones did not bat a gill-flap. Same for the Bridgeport–Flax Pond comparison and the other two pairs of poisoned–pristine sites.

In a paper published in *Science* in 2016, Whitehead and his team showed how the fish had managed to pull this off. From each of the eight locations, they read the genome (the entire length of all the chromosomes, letter by letter) for some fifty mummichog fish. As it turned out, the fish that hailed from the polluted sites all had mutations (rewritings and missing sections of the genetic code) in the genes that code for the AHR proteins, and some also had mutations in the genes of the proteins that AHR interacts with. What is interesting is that many of these mutations varied according to different polluted sites, meaning that evolution had repeatedly, independently, produced PCB tolerance.

They then tested what the effect of these mutations were in living fish and discovered that by and large, the voice of AHR had been muted in pollution-tolerant mummichogs. When exposed to PCBs, AHR no longer switched on as eagerly as in fish from unpolluted environments. So, somehow, the fish had evolved ways to keep their organism developing and running, while essentially removing a few vulnerable cogs and wheels from it.

Presumably this meant some other components had had to be inserted elsewhere or maybe the organism no longer ran as smoothly as it could, but the crucial point is that, thanks to rapid urban evolution, the mummichog is surviving in places where common sense says it shouldn't. "Isn't that remarkable?" Whitehead asks rhetorically. "And that evolved by natural selection over just a few dozen generations!"

PCBs are just one ingredient of the chemical cocktail that we bathe our cities in. Think of road salting. Applying generous quantities of sodium chloride to roads to keep them ice-free in winter is common practice in the colder parts of the world. In the USA alone, a staggering 55 billion pounds of salt are sprayed over its roads each winter. That's a block of table salt about 353 million cubic feet in size per year. No wonder the stuff is pervasive in the environment: salt has been picked up more than a mile away from (and sixty floors above) the roads it is intended to de-ice. Because of this, the water of canals and streams in big cities can be decidedly brackish throughout winter.

For most life forms, all this salinity can be problematic. As we've all learned at school, osmosis causes water to move in the direction of higher salt concentration. That's why, in a salty environment, the cells in the bodies of animals and plants have to work harder to keep pumping the escaping water back in and stop themselves from drying out. And there is another reason why salt isn't good. Chemically, sodium is very similar to potassium. But while potassium is essential for many processes in the cells of animals and plants, the same processes won't work if sodium replaces potassium. And when there's more salt in the environment, sodium insinuating itself into the cell's potassium-powered processes can become a big problem.

Organisms that manage to cope with saline situations have usually evolved mechanisms to counteract the salty onslaught on their cells. And when the environment regularly gets covered in road salt, those same species snap up the places vacated by species that lack such mechanisms. As I mentioned earlier, this is why salt-tolerant beach plants colonize the hard shoulders of major inland roads, pushing out the regular verge verdure. But chances are that the animals and plants that are already there also evolve salt tolerance thanks to road salting.

To test this latter idea, PhD student Kayla Coldsnow and her colleagues of the Rensselaer Polytechnic Institute in Troy, New York, did a laboratory experiment with water fleas (*Daphnia pulex*). Using the same batch of animals, they introduced these tiny freshwater crustaceans into so-called mesocosms (basically large tanks with a real ecosystem in them composed of plankton, plants, clams, snails, and crustaceans). Some of the mesocosms were freshwater, whereas others were brackish (about one-third the salinity of sea water). Yet other ones had intermediate salt concentrations. They left the water fleas to live there for ten weeks (which, in the prolific *Daphnia*, translates to between five and ten generations). At the end, to make sure that any changes were actually genetic, and not some other effect of the saltwater, they took some of the descendants out and cultured those for three more generations in clean lab aquariums with unsalted freshwater. Then they tested each strain for their salt tolerance. As it turned out, the water fleas retained the evolutionary signature of having adapted to salty water. When placed in brackish water with 1.3 grams of salt per liter, the *Daphnia* strains that had lived under moderate salt concentrations in the mesocosm survived well (between 75 and 90 percent made it), whereas the ones previously naïve to salinity experienced only 46 percent survival.

Of course, this is only a laboratory experiment, but there is a good chance that the same kind of evolutionary adaptation to winter salt also takes place in wild animals and plants along major roads, and that we're causing the city flora and fauna of the colder parts of the world to evolve into something akin to a seaside biome.

A seaside biome, yes, but perhaps also a mining town biome. Heavy metals (zinc, copper, lead, for example) are elements that normally are rare in nature. They occur in veins of ore in rocks and under natural circumstances only get into the environment where such a vein hits the surface and is slowly weathered away. All that has changed since humans have discovered the mixed blessings of metals. From the copper axes of yore via the leaded fuels of the twentieth century to today's cobalt, silver, gold, manganese, yttrium, tin, antimony, and gallium that hide in your smartphone . . . Humans are the world's super-accumulators of heavy metals. And much of that accumulation happens in and around cities.

Heavy metals are often toxic, because the molecules tend to attach themselves to enzymes and other proteins, and to DNA, which interferes with the organism's normal functioning. With heavy metals being so rare under natural conditions, most animals and plants have never had a chance to adapt to them, and therefore do not tolerate them well. Enter *Homo sapiens* and its copper slag heaps, leaded petrol, and zinc-coated lampposts and electricity pylons. Suddenly, heavy metals are everywhere. Once again, species have needed to either adapt or disappear.

One species that has adapted is the yellow monkey flower (*Mimulus guttatus*). At the deserted and delightfully named Copperopolis copper mine in California, this species, a widespread wildflower throughout western North America, has evolved ways to deal with high concentrations of copper in the soil thanks to a mutant version of the gene *Multicopper Oxidase*. This mutation, which probably helps to flush copper atoms out of the cell, has become a fixed feature of all the monkey flowers growing on the mine's spoil tips, where they have persisted ever since mining was started 150 years ago.

Something similar appears to have happened to the grasses growing underneath zinc-coated electricity pylons in the UK, where the zinc flaking off the iron structures causes zinc levels in the soil to be up to fifty times higher than normal. In 1988, Sedik Al-Hiyali and colleagues from Liverpool University picked five species of grass from among the feet of pylons that had been erected eighteen to thirty-three years previously, and also from grassland farther away. Then, to gauge the zinc tolerance of these plants they grew them in the lab on zinc-containing soil and measured their root lengths. As it turned out, all five species of grasses growing underneath the pylons grew splendidly, producing roots up to five times longer than the grasses that had been picked from other places.

Not only plants, but even animals in cities have found ways to deal with heavy metals. Between 2000 and 2004, the Russian geneticist N. Yu. Obhukova took it upon himself to travel the length and breadth of Europe and jot down the physical features of almost 9,000 city pigeons. For each pigeon he recorded whether the bird was pale or dark sooty gray—a variability that, in pigeons, is down to genetics. His exercise in pigeonholing revealed that the dark "melanistic" birds, which have much more of the dark pigment melanin in their feathers, were more common in big cities than in less urbanized areas, leaving him to wonder whether this was just the result of genetic mingling with pigeon fanciers' birds, or had another meaning.

In Paris, at the Sorbonne, Marion Chatelain is using the pigeons that so famously blend in with the Parisian zinc-clad cityscape to pick up on Obhukova's hunch. Knowing that melanin binds to metal atoms, she figured that perhaps darker pigeons do better in cities because they are better able to purge their bodies from heavy metal pollutants, such as zinc—simply by transferring the stuff to their feathers. So, she got one of her team members, Lisa Jacquin, to catch some one hundred Paris city pigeons; Chatelain then measured the darkness of their plumage, and housed them under zinc-free conditions in aviaries at her lab, providing each with a leg ring for identification. Then, she plucked two wing feathers from each bird. After one year in the aviary, the feathers had regrown. Chatelain replucked these and did a chemical analysis to see how well the birds had been able to purge their bodies of zinc by storing it in their feathers in the year that they had lived under clean conditions. As it turned out, darker birds had managed to put about 25 percent more zinc in their feathers.

In a follow-up study, Chatelain once again wrangled about a hundred Parisian pigeons to the ground and placed them in her aviaries. Again, she went through her banding and feather-plucking routines, but this time she did not keep the aviaries free of heavy metals. Instead, she divided the birds over aviaries that contained small amounts of lead, zinc, or both, in the drinking water, with two aviaries kept free from pollution as a control. Again, her studies proved that the darker birds stored more zinc *and* lead in their feathers than paler birds, but also showed that surviving chicks who themselves, as well as their parents, had been exposed to lead, exhibit a darker plumage than the ones who had grown up under lead-free conditions. This suggests that paler juveniles had died during the early stage of their life—a sign that there is a real evolutionary advantage to having darker feathers in a polluted environment.

Chatelain's pigeon studies might mean that city pigeons are evolving toward a darker plumage thanks to the detox properties of melanin-laden feathers. But perhaps the story is more complex, because the genes that produce melanin are also involved in stress hormone regulation and the immune system. So the birds' hue and their environment's heavy metal content might not be a simple one-to-one relation, but part of a more complex system: after all, the immune and stress-response systems of urban animals are also taxed differently than those of rural birds. We'll return to this in a later chapter.

So, rather amazingly, some animals and plants can evolve to cope with the most foul and horrid stuff we humans dump in their environment. As we will see later, this is not true for *all* species, for many fail to adapt, and perish. Also the ones that *do* adapt often pay a high price for it. Nonetheless, it is a testament to the power of rapid urban evolution that there are at least some species that manage to keep pace with the chemical pollution that befalls their environment.

12
BRIGHT LIGHTS, BIG CITY

EACH YEAR, NEW YORK CITY COM-
memorates the lives lost during the terrorist attacks of
September 11 with the "Tribute in Light." Eighty-eight xenon
light beams, 8,000 watts each, directed straight up into the
night sky, create two translucent pale blue towers that rise high
above this permanently scarred area in downtown Manhattan.
It's a stunning reminder of the havoc wreaked on that fateful day
in 2001. The installation's producer, the late Michael James
Ahern, once said of it: "It triggers a whole host of feelings and
memories and the things you aspire to, that are without conflict and without aggravation."

And yet, like almost every major undertaking by humans,

even a seemingly pure and ethereal display like this wreaks a havoc of its own. For each year, some tens of thousands of migrating songbirds, which usually fly at night, are trapped in the cage of luminescence. Mid-September being the height of the autumn migration, and the tip of Manhattan a geographical feature that funnels the southbound flyways of multiple species of warblers, each September 11 the memorial display is marred by clouds of confused warblers fluttering from light beam to light beam and uttering a cacophony of alarm calls. Volunteers of the Audubon Society are on site to recover exhausted American Redstarts, Ovenbirds, Black-and-white Warblers, and Northern Parulas, and to advise the display's operators on when to turn the lights off momentarily to allow the birds to regain their bearings and continue their southward migration. Nonetheless, the light display surely leads to bird deaths or, at the very least, to stress and exhaustion on an already taxing trip down south.

A similarly massive, but emotionally very different event of light pollution and its effect on animal mass migration was the finals of the 2016 UEFA European Championship. The football match, between France and Portugal, was staged in the gigantic Stade de France in Paris, on the hot summer night of July 10, 2016. The previous night, for security reasons, ground staff had left the stadium's lights on. Attracted by the gigantic floodlights, thousands upon thousands of moths, mostly the silver Y moth, *Autographa gamma* (named for the bright white Y or γ on its dark-gray speckled forewings), had descended into the empty, cup-shaped stadium. The silver Y is a migratory moth. Each spring, hundreds of millions of them fly north from southern Europe, cruising at altitudes of a few hundred yards above the ground, to benefit from the delayed northern-clime growth spurt in the fields of cabbages, potatoes, and other crops. In some years, there are additional mid-summer migrations across western and northern

Europe and, as moths are wont to, one of these flocks had been lured in by the stadium's lights. Thousands of them were killed by the heat of the lamps, but the rest, dazzled and confused, eventually ended up among the grass of the playing surface where, after the lights were turned off in the morning, they hid throughout the day of the big match.

Then, when evening fell, 80,000 spectators took their seats and the lights were turned back on, causing the sleeping moths to stir. The players' warm-up was already interrupted by clouds of the moths fluttering low across the pitch and by the time of kick-off at 9 p.m., thousands of the insects were zigzagging among the players. Photographs taken that night show annoyed UEFA officials picking moths off each other's dark blue suits, moths blocking the lenses of TV cameras or bunches of them hanging from the goal posts, workers desperately wielding vacuum cleaners to keep the lines on the soccer pitch clear, and, the highlight, Cristiano Ronaldo weeping on the grass after a knee injury in the 24th minute, while a lone *Autographa gamma* sips his teardrops away.

The clouds of birds caught in the Tribute in Light and the masses of moths descending on the Euro 2016 final are just two rather prominent examples of nocturnal animals attracted by artificial lights. The same, in fact, goes on all the time, everywhere, whenever people switch on their incandescent lights, LED displays, gas discharge lamps, or any of the other types of illumination that we have invented to keep out the night-time darkness. For as long as people have done this, this has had the unintended knock-on effect of interfering with the behavior and body clocks of nocturnal animals and even plants.

Moths and other night-active insects are best known for falling prey to light-induced confusion. Light a candle on your porch on a warm summer night and the bugs come flying in from far and wide, circle the flame, searing their wings and eventually doing a kamikaze dive head-first into the boiling-hot candlewax. Scientists still are not completely sure why they do this. Obviously, for the millions of years that insects have been evolving, artificial light wasn't around, so their attraction to light bulbs has got to be the side-effect of some primal behavior, triggered by natural lights. One popular theory posits that night-flying animals use the moon and the stars for navigation. Since these are so far away from earth and move across the sky so slowly, their positions appear static to a cruising insect, and this would allow them to fly in a straight line simply by maintaining a fixed angle with the moon or a bright star. The first insects to come across an artificial light would have treated these the same way. However, ground-bound artificial lights cannot be used as fixed beacons in the same way that astronomical objects can, because they are too close. So, maintaining a fixed angle to a light bulb or a candle flame in fact means drawing closer to it in ever decreasing circles until you snuff in a puff of smoke. In Shakespeare's words, "Thus hath the candle singed the moth."

Whatever the reason, for as long as we humans have lit up the nocturnal environment with our fires, torches, candles, whale-oil lights, and electric lamps, insects have been dying at them by numbers. They die because they burn themselves at close range in the emitted heat or because they are targeted by bats, owls or geckos who have learned that easy pickings are to be had at lampposts. And even if they do not burn or get eaten, simply the time wasted sitting and gazing at the light that they should have spent searching for mates or food could make a hypnotized bug taste defeat in the struggle for life.

If you pay attention to the sheer numbers of insects spinning around street lanterns, caught in the beams of floodlights or forming the accumulated debris caught in the frosted covers of porch lights, you cannot help but wonder exactly how large an impact artificial lighting has on all kinds of animals (insects, mammals, migratory birds, turtles, fishes, snails, amphibians, even plants). They all perform part of their activities under cover of darkness so similarly get confused.

Until recently, science was mute on this question, but for some anecdotal data: 50,000 birds were killed at Warner Robins Air Force Base, Georgia, when they followed landing lights straight into the ground in 1954, and one night in 1981, more than 10,000 birds slammed into floodlit chimneys at an industrial plant near Kingston, Ontario. As for insects, two English entomologists caught more than 50,000 moths at one lamp on the night of August 20, 1949, and once, an estimated 1.5 million mayflies were found dead under the lights on a single German bridge.

For the past fifteen years or so, several scientists have been trying to put more reliable numbers on the impact of "ALAN" (or Artificial Light At Night—light pollution has landed its own acronym in the ecological literature). The German researcher Gerhard Eisenbeis, of the Johannes Gutenberg University in Mainz, for example, has been on the heels of what he calls the "vacuum cleaner effect." As soon as ALAN hits a previously dark area, he writes, "insects are sucked out of habitat areas as if by a vacuum, depleting local populations." For example, the vacuum cleaner effect is probably responsible for the fact that floodlit petrol stations built at remote locations along highways initially attract large numbers of insects, but then, after the first two years or so, the insect numbers visiting the fuel station quickly drop off. By extrapolating from the numbers of insects killed by different types of ALAN, on both moonlit and moonless nights and in different kinds of urban environment, Eisenbeis has come up with an estimate of 100 billion insects killed by ALAN each summer in the whole of Germany—a staggering number, but of the same order of magnitude as the numbers of insects thought to be squished by road traffic.

Though birds are watched much more closely than insects, even avian ALAN-related deaths are hard to gauge. One of the few sets of hard data comes from the Long Point Bird Observatory, on the shore of Lake Erie in Canada. For decades, the observatory has made daily counts of the birds found dead on the lights of the nearby lighthouse out at the very tip of the 15-mile-long Long Point Peninsula. Throughout the 1960s, 1970s, and 1980s, each year there were some 400 kills on the fall migration and about half as many on the spring migration, with occasional peaks of up to 2,000 birds on a single night. But when weaker lights with a much narrower beam were installed in 1989, deaths dropped dramatically to only a few percent of what they had been before.

Kevin Gaston, the urban ecologist whom we already came across in Chapter 5, studying the biodiversity of urban gardens, is another scientist who has taken up the gauntlet and begun series of experiments on the impact of ALAN. I watch him deliver a guest lecture at Leiden University—a friendly but imposing figure, tanned, with the solid features that seem to fit a New York fireman better than an academic ecologist. "People have introduced artificial lighting into the environment on a massive scale and into places and times and forms where it did not occur before," he says. Moreover, he reminds us, "we are moving from narrow light spectra, such as sodium lighting, to a much wider spectrum such as LEDs. It starts to overlap with a whole range of biological sensitivities. It ramifies throughout almost everything."

Given the massive onslaught that ALAN causes, the vacuum cleaner effect, and its rapidly increasing pervasiveness, I put it to Gaston that ALAN may be forcing organisms to evolve some

sort of resistance against attraction by light. But Gaston is dubi-
ous. "This is something these organisms have never seen before,
messing up with those classic daylight cycles of theirs. This has
happened very quickly. I'm not convinced that adaptation is very
easy; some of these light-triggered systems are evolutionarily very
deep-rooted and it may not be possible to adapt to that." But, he
adds, it's not something that has been studied much yet.

He is right there. In fact, all told there are just two articles in
the whole of the accumulated urban biology literature that deal
with evolution in response to ALAN. Quite amazing, really,
given how easy it is to think up an experiment. All you need to do
is select a species of animal which you know is attracted to light,
then catch some individuals of that species in a rural, dark area
and also some in a built-up area with a lot of ALAN, see if you
find differences in how strongly they orient toward light, and
presto! You have your evolutionary experiment.

In fact, that's exactly how Swiss researcher Florian Altermatt
from the University of Zurich approached it. Altermatt, an expert in freshwater biodiversity, is, in his spare time, also a passionate lepidopterist. "My pleasures are—as Vladimir Nabokov
said once—the most intense known to man: writing and butterfly hunting (with the camera ;-)," he proclaims on his website.
He had been using portable mixed mercury vapor lights to attract
and study moths all over Central Europe since he was a kid at
high school, and had long been intrigued by the irresistible lure
of the light on a moth's brain.

So he devised a simple experiment. As his target, he chose the
small ermine moth (*Yponomeuta cagnagella*), so named because of
the regularly spaced black dots on its otherwise pure white
wings—like the multitude of stoats' tail tips on a regal ermine robe.
A sensible choice, because its caterpillars make communal nests in
the European spindle tree, and are very easy to find. Wherever

spindle grows, you can spot the silken nests chock full of their caterpillars. So it was not hard for Altermatt to travel around the city of Basel and also across the border into France, and get fistfuls of baby caterpillars from ten different places. He took care to choose half of those places in urban areas with a lot of artificial light, and the other half in rural places where it was still properly dark at night.

He then put all the caterpillars from the same place in a plastic box with plenty of spindle leaves and left all ten boxes in his lab for the larvae to pupate and develop into moths. When the moths emerged, each received a mark to distinguish dark-sky from urban moths and then he released all of them, 320 rural moths and 728 urban moths, in one go into a dark room with a fluorescent light trap at one end. The idea, of course, was to see how many of each type would end up at the light. The results, published in 2016 in the journal *Biology Letters*, showed a clear signature of urban evolution: whereas 40 percent of the country moths flew straight to the light, only about a quarter of the city moths did so—the rest staying put where they had been released.

This simple experiment, on a random moth, shows exactly what would be expected if ALAN had been purging flight-to-light genes from the urban population. Is this a common situation in more insects? Are all urban insects perhaps evolving the ability to withstand the lure of the lamp? We won't know until Altermatt's experiments are repeated with other species and on a considerably larger scale.

Elsewhere in Central Europe, a further twist to urban evolution in response to light pollution has been revealed. In the late 1990s, Astrid Heiling, an arachnologist at the University of Vienna, studied the urban spider *Larinioides sclopetarius*. Commonly known as the bridge spider, you can find it building its webs over water (and, indeed, often on bridges) in both urban and rural settings around the globe on the northern hemisphere. But Heiling did not study them all over the northern hemisphere. Instead, she focused on a single 60-yard-long pedestrian bridge across the Danube Canal in the heart of Vienna.

Here, the spiders built their webs in the open spaces above the handrails. There were four handrails along the bridge, Heiling explains in the article she published in *Behavioral Ecology and Sociobiology*. Two were lit with fluorescent tubes, the other two were dark. For an entire summer, Heiling walked up and down this footbridge, doing daily censuses of the bridge spiders at the handrails. Ignoring the curious glances of passersby, she accumulated notes that revealed to her that the spiders predominantly built their webs near the lights: in early autumn, the illuminated handrails supported almost 1,500 fat spiders, on average four per square yard, their webs sometimes even overlapping. The dark handrails, on the other hand, housed only a few hundred. And what's more, the spiders in the artificially lit "habitat" caught up to four times more prey in their webs than the ones that stayed in the dark—not surprising, given insects' propensity to fly to light.

Unlike insects, spiders normally are not lured in by light at all. Quite the contrary: they tend to flee from artificial light and hide in dark corners instead. So what Heiling wanted to know was: how did the spiders manage to seek out the best visited web sites? Did they just move around until they found a place with good prey traffic, that is, near the lights? Or had they perhaps evolved an attraction to the light? To put this idea to the test, she devised an experiment. She caught spiders and, in her lab, placed them in tanks with a dark end and a lit end. To filter out any effects of the spiders learning, rather than evolving, the blessings of light, she did the same with adult spiders that she had hand-reared in the dark in the lab. Nearly all the spiders, the ones that came straight from the lit handrails, as well as the inexperienced ones that had grown up in the lab and had never seen a light bulb, went straight for the lit end of their tanks and constructed their webs there.

Unfortunately, Heiling never carried out the same experiments with bridge spiders from a non-urban place without any light pollution. Still, while not completely water-tight, her results give the impression that the spiders have evolved their attraction to light to exploit the droves of winged insect-snacks attracted to artificial lighting.

The genetically-based ALAN attraction and repulsion, respectively, that Heiling and Altermatt have uncovered, are crying out for follow-up studies. With artificial lights making such astronomical numbers of victims among nocturnal animals, and probably more subtle effects on the daily lives of all organisms, it seems that resistance to the innate lure of the light must be evolving all over the place. And yet, until now only very few biologists seem to have been interested in revealing the workings of this kind of urban evolution. It's time for more researchers to see the light!

13

BUT IS IT REALLY
EVOLUTION?

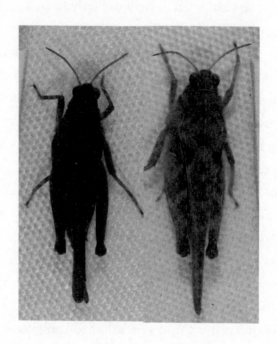

IN 2016, *THE NEW YORK TIMES* ASKED me to write an article about urban evolution. After the piece came out, I received dozens of emails from intrigued readers, many of whom shared with me their own observations of urban wildlife. A Mr. Spanier, who lived for many years in Santiago, Chile, wrote about the stray dogs there and how an out-of-town visitor, riding as a passenger in his car, suddenly exclaimed, "That dog back there looked *both ways* before stepping into the road!" Could that also be down to evolution, he wondered.

You may have asked yourself the same question while reading

the previous chapters. Or you may have doubted that the examples I have given can truly be called evolution. After all, with a few exceptions (which I'll save for later), we are not talking about the evolution of entirely new forms of life. Mostly, urban evolution is more subtle than that—not surprising, because it takes place within such brief time frames. The urban ermine moths of Basel that are less attracted to light are only so by a small degree compared with their rural relatives. The investment in heavy-weight seeds by the urban hawksbeard in Montpellier is only a fraction greater than outside the city. Measurable and statistically significant, yes, but still only rather minuscule. It's not as if, when shown an urban and a rural hawksbeard plant, you are blown away by the difference. To the untrained eye, they would still look precisely the same.

What's more, the genetic tweaks that allow urban evolution to take place are often not new. The dark feathers that aid urban pigeons in dealing with heavy metals, for example, are darkened thanks to a mutant plumage color gene that also exists in wild rock pigeons and was already there long before these birds were tamed by man, escaped, and became city birds. And the genes that permit plants to grow on urban soil polluted with heavy metals are exactly the same ones that have already floated around in populations of these plants for thousands of years, maintained because they occasionally save a stand that grows on some talus slope rich in copper or zinc minerals.

In fact, ever since geneticists began screening chromosomes, they have been surprised by how genetically variable most species in nature are. Pick a gene, any gene, from a wild species—say, a gene that affects leg length in the *Anolis* lizards in the cities of Puerto Rico. Usually, such a gene will consist of a sequence of several thousands of DNA "letters" of genetic code. This genetic code specifies the exact structure and shape of a protein. In the case of

Anolis's leg-length gene, this protein may, for example, let cells in the embryo divide at the rate and directions required to produce a limb-like outgrowth.

The thing is, the genetic code for a gene is very rarely exactly the same across all individuals of a species. It could very well be that, if you were to read the code for the lizard-leg gene in a thousand *Anolis* lizards from a Puerto Rican city, you'd find thirty or forty different versions of that same code. Most of those would differ from one another in minor details only: a changed letter here and there, perhaps a short bit of DNA deleted or duplicated—the products of copying errors made in the loins of some ancestral lizard many generations before. And, in most cases, these variants would all still do the exact same thing, that is, produce a lizard leg in a lizard embryo—hence their unhindered passing down the generations. But, it could be that one variant produces an almost imperceptibly slenderer leg, or a slightly stubbier one; or one that begins growing just a little later in the animal's development.

This palette of subtly different gene variants, most of which are so similar that they are inconsequential for a species' evolution most of the time, is called "standing genetic variation." And it is this palette that evolution dabs its brushes in to create new works of urban art. For when the environment changes and suddenly offers a new, previously nonexistent benefit for *Anolis* lizards with slightly longer legs, those gene variants already able to do the job are right there in the population, ready to rise to the occasion and yield to the opportunities of natural selection.

Hence, standing genetic variation is a species' evolutionary capital. It encapsulates the ability of a species to dip into its genetic savings and immediately come up with any combination of genes that a changed environment requires. That is why urban evolution can proceed so rapidly: the animals and plants that need to adapt to whatever new feature humans release in their urban environment do not need to wait for the right mutations to come along. Mostly, the necessary gene variants are already there, waiting in the wings of the standing genetic variation. It only takes natural selection to bring them out into the limelight, and give them a chance to shine.

Evolution employing pre-existing gene variants is what biologists call "soft selection." But urban evolution could also make use of new mutants that arise there and then ("hard selection"). To distinguish soft and hard selection, geneticists do a close-reading of the genetic code involved in an urban adaptation. In the PCB-tolerant mummichog fish, for example, Andrew Whitehead discovered that different genes are involved in different harbors along the US east coast. And even within the same harbor, the same gene variant that affords protection against PCBs may, in different mummichog fish, be flanked by different genetic codes on either side. These are all tell-tale signs that the PCB-tolerant gene variant already arose long ago and has since become detached from its neighbors due to the frequent chromosome breaking and crossing-over that occur during reproduction. So, the mummichog PCB tolerance clearly evolved via soft selection, using only its pre-existing standing genetic variation.

But as we saw in a previous chapter, it's a different story for the peppered moth in England. It adapted to soot-covered trees by virtue of a mutation that took place in the *cortex* gene right at the onset of the industrial revolution. Throughout all of England, the dark-colored moths carry the same variant of *cortex* and the gene's neighboring regions are also identical. This is a clear signal of hard selection: the new *cortex* gene gave such an advantage that it swept through the population like wildfire, so fast and pervasive that it dragged its neighboring genes with it, before these found the time to be decoupled by chromosome breakage.

So, yes, rapid urban change may be subtle. It may also employ genes that are already floating around in the species. Nonetheless, it really is evolution, pure and simple.

One of the readers of my *New York Times* piece was not convinced. Cornelius Hunter, a blogger for the website *Darwin's God*, used my exposition of urban evolution as a peg for a fine piece of creationist writing, in which he stated, "Adaptation and evolution are two very different things. Biological adaptation relies on the preexistence of [...] genes, alleles, proteins, [...] and so forth. Evolution, on the other hand, is [...] the origin of all those things."

In other words, Hunter the creationist made a distinction between soft selection, which he saw as an inevitable physical process, building on already existing materials, and the origin of something entirely new, new genes and new "kinds of organisms." Only the latter, in Hunter's opinion, deserves to be called evolution (and, needless to say, to his mind cannot exist).

It's amusing to see how creationism, in the face of ever-improving evolutionary knowledge, keeps moving the goalposts about what counts as evolution. Fortunately, this is not a matter of opinion. Biologists have a very clear definition of what evolution is, namely the change, over time, of the frequencies of gene variants. The *origin* of those gene variants is, contrary to what the *Darwin's God* blog would like its readers to believe, *not* evolution. Instead, it is chemistry—errors made in the stitching together of DNA from molecular building blocks in the cell. But the natural selection that causes these chemically originated gene variants to become common or rare, that *is* the stuff of evolution. Over longer periods of time such small, short-term evolutionary steps coalesce into bigger evolutionary changes for many different genes, eventually leading to entirely new species.

But even if you're not a rabid creationist, and perfectly able to see that the urban hawksbeards, mummichog, anoles, bridge spiders, ermine moths, and all the other creatures that have peppered these pages, did, in fact, evolve, there may be room for doubt. After all, for a trait to actually evolve, it is necessary that it is genetic—encoded in the organism's DNA. And biologists who see a change in an urban organism's looks or behavior do not always have firm evidence that this is the case.

The color and color patterns of animals, for example, are usually genetic. Our own bodies are testimony to that: the color of our hair, skin, and eyes is determined by the genes we inherit from our parents. But we also know that skin can be tanned and hair can be bleached by sunlight, and eye color sometimes changes as we age. So, genes are not everything. For most aspects of our appearance, both nature and nurture come into play. This is true for animals, too. For example, when *Tetrix* ground hoppers (a kind of drab, camouflaged grasshopper) grow up on light-colored sand, their body color is lighter than when the same animals would grow up on dark sand. This is a form of "plasticity": with the same DNA, you can get different outcomes. So, it's easy to imagine that when, in a certain species of animal or plant, you find a color difference between the ones living in the city and the ones living outside, you might mistakenly think this is urban evolution, whereas in fact, there is only plasticity.

Behavior is especially tricky. If some species of bird, for example, behaves more boldly in the city than in the countryside, this does not necessarily mean that the city gene pool is rich in genes for bold behavior. They could simply have *learned* that boldness pays off in an environment where there is food to be snatched and fear for predators is unwarranted. Or, the reverse, the countryside birds may have learned that out in the wild nature, one had better be circumspect, whereas the city birds have grown more naïve.

We know that in many animals, behavior can be genetically determined, but we also know that it can be learned and passed on from one individual to the next by instruction and imitation. To figure out what the relative contribution of each is, you'd have to do complicated experiments involving a lot of rearing and crossing and breeding—something that is not always practicable. The reason that Astrid Heiling in the previous chapter took the trouble to hatch bridge spider eggs and hand-rear the spiderlings in the lab was to make sure that the spider's attraction to light was innate, and not a habit they had picked up while growing up in an artificially lit environment. And even if an urban behavior is entirely down to an animal's ability to learn, this is not to say that evolution may not eventually kick in. If boldness is beneficial, for example, then learning to be bold may, over many generations, be replaced by *genes* that cause an animal to be bold from birth—hard-wiring the useful behavior that otherwise would need to be built up slowly during life.

Then there is epigenetics. I apologize for littering this chapter with new terms, but I promise that epigenetics is the last one. It might be an important one. We don't really know, because epigenetics is still such a new thing in evolutionary research.

The meaning of the term "epigenetics" was only cast in scientific stone at a conference in Cold Spring Harbor Laboratory in 2008. It denotes a change in some characteristic of an animal or plant that is the result of "changes in a chromosome without alterations in the DNA sequence." Now that may sound strange because chromosomes are made of DNA, right? Well, yes and no. Chromosomes contain the DNA, but they are much more: they also contain proteins and other molecules that package the DNA like bubble wrap. And only when the packaging is peeled off to reveal the naked DNA, can a gene do its work. As it turns out, some of this packaging material can be added or removed during an animal or plant's life to muffle or amplify a gene's voice, as it were.

What's more, the offspring can sometimes inherit certain types of this packaging conformation. So, if the toils of life place high demands on a certain gene, then some of the packaging may be removed, making this gene churn out protein at a higher rate. The offspring may then inherit DNA from their parents with the packaging pre-removed, to give them a better start in life. That way, epigenetics can enhance or suppress a certain gene for several generations on end. You can probably imagine that this could lead an urban evolutionary biologist astray. If some aspect of an organism is beneficial and genetic, then an increase in it in a city environment would be taken as a clear sign that it is evolving. And yet it may be that the DNA is actually unchanged and that the "evolution" is really only epigenetics at play.

The salt tolerance of Kayla Coldsnow's *Daphnia* water fleas, for example, might be down to epigenetics. It's known that when these animals are exposed to toxic chemicals, they turn on genes that help detoxify their bodies. And it seems that their offspring are born with the epigenetic switch already in the "on" position. If the same thing is true for salt, then this could explain the very fast adaptation to road salting that Coldsnow discovered. Only finding and sequencing the genetic codes of the genes involved would definitively answer this question.

As Kevin Gaston says in an overview article, almost no study in urban evolution has distinguished between genetic adaptation and epigenetic effects. "Addressing these limitations will be a major challenge for urban ecology in the future." At the moment, most experts think rapid urban evolution is still mostly due to actual changes in the DNA, rather than epigenetics, but few people dare to state this with any great confidence, and the next few years may well prove that epigenetics is a force to be reckoned with.

I hope you'll forgive me this short exposé on epigenetics, plasticity, soft and hard selection, and all the other intricacies surrounding urban evolution. With a process that is so subtle, and yet so important for the survival of biodiversity in our urbanized future world, it's crucial to be well versed in modern evolutionary biology. Now you are ready to be taken to the next level of urban evolution. Enter: the Red Queen!

III.

CITY
ENCOUNTERS

"Well, in our country," said Alice, still panting a little, "you'd generally get to somewhere else—if you run very fast for a long time, as we've been doing."

"A slow sort of country!" said the Queen. "Now, here, you see, it takes all the running you can do, to keep in the same place. If you want to get somewhere else, you must run at least twice as fast as that!"

LEWIS CARROLL, *Through the Looking-Glass* (1871)

14
CLOSE URBAN
ENCOUNTERS

You PROBABLY KNOW THOSE SCENES
of killer whales launching themselves onto Argentinian
beaches to snatch up unsuspecting seals. Dramatic footage of
the bulky black-and-white giants emerging from the surf and
grabbing seals as if they are cookies on a counter has featured in
countless nature documentaries. If you hold on to that image and
scale it down, then you have something akin to what French biologists in the city of Albi began witnessing in 2011.

Albi is a small city in southern France, the capital of the department of Tarn, named after the river that slowly meanders through
Albi's medieval, UNESCO-registered center. In the heart of the

city, the Tarn is spanned by the Pont Vieux, a bridge that connects two heavily built-up districts on either side. Like urban areas everywhere, these city quarters are home to the customary flocks of feral pigeons. Here, though, the pigeons have more immediate concerns than the accumulation of lead and zinc in their feathers that we saw earlier in this book. For every day, when groups of pigeons gather on a gravel island underneath the Pont Vieux to bathe and preen, they are the target of what French biologists Julien Cucherousset and Frédéric Santoul call "freshwater killer whales."

The killer whales in question, Cucherousset and Santoul explain in an article in the journal *PLoS ONE*, are European catfish (*Silurus glanis*). The continent's largest freshwater fish, they can easily reach a length of five feet, and even individuals of over six and a half feet are occasionally reported. Catfish are native to eastern Europe and western Asia, but they did not occur in western Europe until people began releasing them for sport. The river Tarn was first stocked with catfish in 1983 by local angling societies. The fish did well, expanding fast on a diet of smaller fish, crayfish, worms, and molluscs that live in the river's muddy bottom. But at some point, the urban catfish of Albi began doing something that no other catfish had ever been seen doing before: launching themselves out of the water, grabbing bathing pigeons by their feet, dragging them under water and swallowing them, metal-laden feathers and all.

For a whole summer, Cucherousset, Santoul, and their team of students took turns observing this spectacle from the Pont Vieux, capturing a total of 72 hours of fish-pigeon encounters. Their videos, shot straight from above, make for nail-biting viewing. We see a flock of pigeons frolicking on the gravel beach, dipping their beaks into the river to drink, and happily flapping their wings to spray themselves with river water—seemingly oblivious to any danger. Meanwhile, a menacing dark shape in the water looms closer and closer. As the catfish approaches the shore where the birds are splashing about (you have to imagine the musical soundtrack), it raises the long whisker-like barbels around its mouth to detect its prey's vibrations more precisely. It singles out one bird that is standing with its feet just in the water and then, with a few wild tail swings, launches itself onto the beach, grabs the pigeon by a foot and quickly retreats back into its watery habitat, taking the desperately flapping bird with it, and swallowing it in a few large gulps of its gaping mouth. The other pigeons take off in momentary consternation, but then return to their bathing spot as if nothing had happened. Soon, a second catfish attempts the same trick on another pigeon. In all, the researchers filmed fifty-four attacks, of which about a third were successful. A chemical analysis of the catfish, the pigeons, and several other prey animals (smaller fish and crayfish) proved that the pigeons make up around a quarter of the catfish diet, with some catfish individuals getting nearly half their nutrients just from eating birds.

Eating birds. *Beaching themselves* to catch them, for crying out loud! Let this sink in for a second. According to the handbooks, catfish are supposed to feed on fish and aquatic invertebrates that they rustle up from the mud. That is their catfish niche. They never evolved to launch their bulky bodies onto shore to pull winged animals down in mid-takeoff. And yet, here we are: humans brought rock pigeons and catfish into their cities, introduced them to each other, and thus created an ecological opportunity that never existed before.

In the previous chapters, we have seen animals and plants adapt to the physical features of the city: its countenance of glass and steel, the deadly pulse of its traffic-filled veins, the luminous mantle of artificial light that it cloaks itself in, and the trickles of foul chemicals oozing from its pores. All of these evolutionary events are the result of one particular type of close encounter between the urban environment and wild animals and plants. Let's call those close encounters of the first kind: the evolving organism is the moving part, the physical feature is static.

For example, in 2017, Sarah Diamond of Case Western Reserve University discovered that acorn ants (*Temnothorax curvispinosus*) adapt to the urban heat island. A colony of speck-like acorn ants fits inside a single acorn, and since oak trees occur both inside and outside the city, Diamond and her colleagues could investigate the ants' tolerance of higher temperatures simply by picking up acorns with ants inside and moving them to warmer or colder places. In doing so, they discovered that the city ants could stand the heat a bit better than their rural formicine relatives and they also proved that this difference is partly genetic. Once again, a very nice example of urban evolution, similar to many others we have seen before. But it is important to remember that this is a one-way adaptation. The heat island itself is completely unaffected by the fact that this animal has adapted to it.

Obviously, there will be no feedback between the acorn ant's evolving ability to stand the urban heat and the heat island itself. But, the same is not necessarily true for close encounters of the second kind, such as the unfortunate interaction between catfish and pigeons in Albi, France. Here, a situation is created where both sides of the interaction could adapt to one another. The catfish could evolve to improve their bird-grabbing abilities, whereas the pigeons could evolve a greater wariness around water. At the moment there is no evidence yet that either species is evolving. Still, the scene is set for such two-way evolution.

Being either one-way or two-way evolution is one important distinction between the close urban encounters of the first and second kind. But there is more. The first kind could, in principle, come to a standstill. As soon as, say, the mummichog of Bridge-port have reached peak tolerance to PCBs, this evolutionary process is completed. The newly evolved, PCB-adapted mum-michog will continue living in its polluted waters for as long as it likes. With the second kind, such an evolutionary backwater may never be reached. If the pigeons evolve a more circumspect per-sonality, this may result in the catfish evolving a greater speed of attack, leading to the pigeons evolving a quicker flight response, which could then precipitate a greater sensitivity of the catfish whiskers, and so on, and so forth. It is not likely that this will ac-tually happen, if only because the catfish does not depend en-tirely on pigeons for food, and because there may be catfish-free places for the pigeons to do their daily ablutions. But, in theory, we could see endless cycles of mutual urban evolution of catfish offense and pigeon defense.

The endlessness of evolutionary adaptation when the thing that one adapts to is not a physical feature but another organism that can itself evolve is what makes this second type of evolution so powerful. The evolution of one partner fuels the evolution of the other partner, and the net effect is that they remain locked in an ecological interaction with one another, like two countries forced to engage in a perpetual arms race simply to prevent one from overrunning the other. It is for this reason that evolutionary biologists call this type of antagonistic adaptation "The Red Queen," after the character in *Through the Looking Glass*, who told Alice, "Now, here, you see, it takes all the running you can do, to keep in the same place."

But the evolving partners do not even have to be archenemies in order to influence each other's evolution. All animal and plant species in the urban environment form knots in a gigantic, con-stantly changing tapestry of ecological interactions. Sure, there are plenty of species in this vast urban ecosystem that are at each other's throats. But there are also many that simply jostle for space in the cracks of the pavement, or that help each other gain a foothold. Think of the sparrows nesting in the ivy growing on buildings, or the springtails finding shelter among the succulent plants on green roofs. Whatever the interaction, chances are that if one species evolves, this will also affect some of the other species that are tied with it in the urban ecological web. No species, after all, is an island.

As we have seen before, cities are like mad scientists, creating their own crazy ecological concoctions by throwing all kinds of native and foreign elements into the urban melting pot. Our gardens, balconies, and parks are stocked with plants from all over the world, which then provide food for a motley crew of animals from all continents. In Paris, Indian ring-necked parakeets are eating seeds of the black locust trees from North America. In Malaysian cities, European rock pigeons are ripping out the flower buds of Chinese hibiscus bushes planted along the pavements. In Perth, the Indian northern palm squirrel was released in 1898 and since then has maintained a healthy population thanks to the abundant fruits of African date palms and other exotic palm trees in the city.

The urban loom weaves food webs from weft and warp that are thrown together by chance, linking species in new and exciting patterns. Since such ecological interactions are marriages of convenience, rather than matches made in heaven, the species thus linked may evolve adaptations to dealing with their new ecological counterparts. Some of the finest examples of this come from plant-eating animals, so-called herbivores. In Florida, for example, one can find the native soapberry bug, *Jadera haematoloma*. This insect feeds on the seeds of the (also native) balloon vine, *Cardiospermum corindum*. The balloon vine is so named because it carries its tiny seeds inside a green bubble of two centimeters in diameter, and the soapberry bug plies its nearly 9-mm-long snout to pierce that bubble and *juuuust* manage to get to the seeds in the center.

Around 1955, the Taiwanese raintree (*Koelreuteria elegans*) was one of the exotic trees that the Florida parks authorities began planting in parks and along roadsides. The raintree is related to the balloon vine, but has a much smaller and flatter seed capsule. At some point after its arrival, soapberry bugs began eating the seeds of the raintree as well. And, as Scott Carroll of the University of California discovered in the 1990s, as a consequence, the raintree-dwelling bugs evolved, almost to the extent of becoming a separate species. Just forty years after the raintree began to become a common sight along Florida's streets, the bugs living on them lay more, but smaller, eggs, they develop faster, and they are attracted by the scent of raintrees, not balloon vine. But the most eye-catching difference is in their snouts, which, in the bugs on raintrees, are shorter: only about 6.5 to 7 mm long. Shorter than in their balloon vine–dwelling ancestors (in fact, too short to be any good on a balloon vine pod), but long enough to reach the seeds inside the much smaller seed capsules of the raintree. What's more, Carroll showed that all these differences between the old and the new version of the soapberry bug are coded in their DNA.

In 2005, Carroll announced a cute further twist to this story. For in Australia, the same sequence of events unfolded itself, but then, in good Notogean fashion, upside down. In Brisbane, another species of soapberry bug, *Leptocoris tagalicus*, lived mainly on the native woolly rambutan (*Alectryon tomentosus*), until the American balloon vine was introduced there, eventually becoming a nationwide pest around 1960. Roundabout that year, the Australian soapberry bugs, sufficiently provoked by the balloon vine's abundance, hopped onto it. Carroll measured the snout lengths of pinned *Leptocoris* in Australian natural history museums and discovered that before 1965, they all had short snouts, whereas after that year, longer-snouted individuals began to appear. These pre-1965 longsnouters presumably were the first ones to have colonized and adapted to balloon vine. Today, as Carroll discovered, the *Leptocoris tagalicus* bugs on balloon vine have snouts a tad longer than those on the woolly rambutan, long enough better to reach into balloon vine seeds.

The Pinocchio-like growing and shrinking of soapberry bugs' noses is a textbook example of herbivores evolving after they jump to a new, introduced food plant. Many of these cases come, of course, from agriculture, where such a shift onto a crop usually means the dreaded appearance of a new pest. In the Hudson River Valley of the US, for example, the native hawthorn fly spawned a new species that, over the past few hundred years, adapted to the apple after it was introduced there by settlers from Europe. The apple maggot fly (*Rhagoletis pomonella*) has by now become so different from the hawthorn fly that many consider it a separate species. And in Europe, *Ostrinia scapulalis*, a native moth burrowing in the stems of the native plant mugwort, also gave rise to a new species, *Ostrinia nubilalis* (the European corn borer) when maize was brought to Europe from America around 1500. Within that half millenium, the European corn borer has evolved lots of maize-specific adaptations, including one particularly delightful one. In late summer, the caterpillars chewing away inside the plant stalks go into so-called "diapause," basically an extended holiday before the hard work of metamorphosis begins. But while *Ostrinia scapulalis* caterpillars install themselves somewhere in the middle of their plant's stalk, those of *Ostrinia nubilalis* first burrow down the corn stalk until close to ground level. Why? Think of the decades of natural selection caused by the late sum-

mer onslaught of the corn harvesting combine, and you'll probably know the answer!

Scientists have now accumulated evidence of dozens of plant-
eaters colonizing an exotic plant and then evolving in ways similar to the soapberry bug, the apple maggot fly, and the European
corn borer. My students and I also contributed: we discovered
that in the north of the Netherlands, the leaf beetle *Gonioctena
quinquepunctata* moved from the native rowan tree (*Sorbus aucu-
paria*) to the notorious invasive American black cherry (*Prunus
serotina*), a shift that is very recent (it happened around 1990), but
that already shows up as changes in several of the beetle's genes.

Plant-eating animals adapting to a new food plant form one
aspect of the Red Queen game. The reverse, plants adapting to
new plant-eating animals, is another. Cordgrass, for example, is a
tough type of grass that grows on coastal marshes along all coasts
of the Atlantic Ocean. With such a reputation and an equally
awe-inspiring scientific name, *Spartina* has been the plant of choice
for constructing bull's eyes on archery targets and has also been
sturdy enough to hitch rides with humans across the globe to salt
marshes on all the world's shores. Smooth cordgrass, *Spartina
alterniflora*, for example, is a species from the North American
east coast, but humans have accidentally carried it to the continent's
west coast where it now thrives in places as diverse as Washington state's pristine Willapa Bay (where it has lived since around
1900) and the urbanized shores of San Francisco Bay (year of
arrival: 1970).

But being in an urban or a rural environment is not the only
difference between these two new homes for smooth cordgrass.
For in Willapa Bay, the grass is blissfully unaffected by any insect pests, whereas in Frisco, it finds its leaves sucked dry by the
plant-hopper *Prokelisia marginata*, an east coast insect that is as

non-native to the city as people who say "Frisco." Two researchers, Curtis Daehler and Donald Strong, studied in a greenhouse
whether this difference in herbivory has made the cordgrass in
both places evolve in different ways. Sure enough, they found that
the San Francisco plants, when attacked by the plant-hoppers, only
lost about 20 percent of their leaves and happily lived on, whereas
the ones from three states up, evolutionarily unprepared for the
insect, suffered 80 percent leaf loss and nearly half of them succumbed. Apparently, the two cordgrass colonies had evolved opposite pest resistance, perhaps having to do with the chemicals
they use to make their leaves unpalatable.

In a newly discovered twist, some of the chemicals that plants
use to defend themselves against plant-eating insects pass through
human hands, to be then used as natural insecticide by birds to
fumigate their nests with. Okay, read that sentence once more.
Sounds intriguingly convoluted, right? Well, imagine how intrigued the Mexican ornithologist Monserrat Suárez-Rodriguez
was when, in 2011, she began discovering discarded cigarette
butts in the nests of house sparrows and house finches at the
Campus of the National University of Mexico in Mexico City.
Discarded cigarette butts are a ubiquitous urban eyesore all over
the world. We are all taught at school not to throw any litter
on the street, but smokers appear to have collectively decided
that this obviously does not apply to the cool photogenic finger
flick with which they rid themselves of butt leftovers. Globally,
5 trillion (yes, that's a five with twelve zeroes) filter cigarettes are
smoked per year, and many of those filters end up in the environment, where they take several years to degrade. No wonder, perhaps, that the Mexican urban birds could not avoid mixing them
in with their nest material. Suárez-Rodriguez found up to forty-
eight butts per nest. Basically, these birds were brooding in an
ashtray.

But was it really an accidental inclusion into the birds' nest material, or was there perhaps something else going on, Suárez-Rodriguez wondered. After all, some birds are known to incorporate green plants in their nests, because the chemical compounds in the leaves keep mites, fleas, and lice at bay. And since cigarettes are made of leaves of the tobacco plant, whose main anti-insect agent is nicotine, perhaps the campus birds benefited indirectly from the human penchant for that same chemical compound. To test this, Suárez-Rodriguez and her colleagues measured the amount of cigarette butts in some sixty nests and also counted the numbers of mites in them. They found a beautiful negative relationship: more butts meant fewer mites, while birds that refused to turn their nests into smokers' dens paid a dear price for their cleanliness. They had to share their nests with up to a hundred blood-sucking mites, whereas the nests with more than ten grams of cigarette material were virtually free of mites.

Unfortunately, we don't know yet what lies at the root of these bird equivalents of the bug bomb. It could be that the birds are sensing nicotine in the butts and are treating them as if they were the fresh plant leaves they would otherwise have integrated with their nest material. It could also be that, over successive generations, the birds learned that nests with more butts were more comfy. Or, the behavior could have a genetic basis and be a newly evolved defense of bird against bug. If so, the next task for the Mexican researchers would be to see if urban bird-nest mites are evolving a nicotine resistance.

Granted, what I have shown you here are not really full cycles of Red Queen evolution. We've seen herbivores adapting to plants that humans have brought into their environment, and we've seen *other* plants adapting to herbivores that feed on them thanks to human intervention. We have even seen birds controlling their parasitic mites with insecticides derived from plants and made available by urbanites' smoking habits. What we haven't got yet are good examples of the same ecological interaction going through successive evolutionary cycles of attack, defense, counterattack, and counter-defense. This probably has to do more with the fact that most biologists are either zoologists or botanists (so they will view the interaction either from the plant's or the herbivore's viewpoint), than with a rarity of such events. We see fragments of such cycles in different species, so it is quite likely that there are new urban ecological relations that are actually going through such tit-for-tat adaptation right here, right now.

15
SELF-DOMESTICATION

A CRUMBLING CONCRETE WALL, A ramp, and a vast expanse of tarmac on which identical silvery-gray sedans are slowly circling and zigzagging between traffic cones. It does not seem like much, but to urban biologists the Kadan driving school in the Japanese city of Sendai is hallowed ground. The four of us (biology students Minoru Chiba and Yawara Takeda, biologist Iva Njunjić, and I) have been sitting on that crumbling wall now for several hours, hoping to observe what this place is famous for.

It is here that, in 1975, the local carrion crows (*Corvus corone*) discovered how to use cars as nutcrackers. The crows have a predilection for the Japanese walnut (*Juglans ailantifolia*), which grows abundantly in the city. The pretty nuts (a bit smaller than

commercial walnuts, and with a handsome heart-shaped interior)
are too tough for the crows to crack with their beaks, so for time
immemorial they have been dropping them from the air onto
rocks to open them. Everywhere in the city, you find parking lots
strewn with the empty nutshells: the crows either drop them in
flight or carry them to the tops of adjoining buildings and then
throw them over the edge onto the asphalt below.

But all this flying up and down is tiring, and sometimes the
nuts need to be dropped repeatedly before they split. So, at some
point, these crows came up with a better idea. They would drop
nuts among the wheels of slow-driving cars, and pick up the flesh
after the car had passed. The behavior started at the Kadan driving school, where there are plenty of slow-moving cars, was copied by other crows, and so spread to other places in the city where
slow-moving giant nutcrackers were common, such as near sharp
bends in the road, and at intersections. At such places, rather than
dropping the nuts from above, the crows would station themselves
by the roadside and place them more accurately on the road. Since
then, the fad has also turned up in other cities in Japan.

A zoologist of Sendai's Tohoku University, Yoshiaki Nihei,
made a detailed study of the behavior. He observed how the crows
would wait near a traffic light, wait for it to turn red, then step in
front of the cars, place their nuts, and hop back to the curb to
wait for the lights to turn. When the traffic had passed, they would
return onto the tarmac to retrieve their quarry. His work revealed
the crows' finesse in handling their "tool." For example, the birds
would sometimes move a walnut a few centimeters if it took too
long for it to be hit by a wheel. In one case, he even saw how a
crow would walk into the path of an oncoming car, forcing it to
brake, and then quickly toss a nut in front of its wheels.

These fascinating observations languished in relatively obscure
Japanese scientific papers until 1997. That year, the BBC came to

Sendai to film the crows for David Attenborough's series *The Life of Birds*. Sir David's voice-over made them an instant hit: "They station themselves at pedestrian crossings . . . Wait for the lights to stop the traffic. Then, collect your cracked nut in safety!"

So, finding ourselves in this city with its famous urban crows, my merry band devoted today to view them for ourselves. Minoru and Yawara tell us that the crows' trick is well known in town. In fact, it is a favorite pastime to throw the crows nuts and watch them perform. So, with a bag of walnuts brought all the way from the Netherlands, we try our luck. But the crows are not cooperating. We have already spent the whole morning at traffic lights at intersections, stupidly waiting on canvas folding chairs at the mercy of the surprised stares of endless motorists, but so far, in vain. And we have now ended up at the reputed epicenter, the Kadan driving school. It is getting hot and we're hungry and tired. With glazed-over eyes, we stare at the heaps of nuts we have laid out at various positions on the school's test range. The school's students carefully avoid them, and the crows fly over without even looking down. This is what urban fieldwork is like.

Perhaps, Minoru and Yawara finally admit, it is too early in the year. The nuts are not ripe yet, the young birds have just fledged, and groups of crows are marauding the city to feast on other things, like ripe mulberries that are in abundance everywhere. I sigh and stare a bit more. Then, I hear a cracking noise behind me. I turn around to see that Iva has begun eating our stock of walnuts. She looks at me defiantly. "What? They're not going to come anyway!"

Carrion crows do not only occur in Japan. They also exist in western Europe, where you can similarly find plenty of cars, pedestrian crossings, and walnuts. And yet, carrion crows in Europe somehow never learned to exploit human automobile traffic in the Rube-Goldbergian way that they do in Japan. That is not to say that humans in Europe are safe from having their behavior manipulated by birds, as demonstrated to us for nearly a century by the famous (and annoying) milk bottle top-opening skills of tits—lively songbirds with a handsome pattern of yellow, black, and blue (the blue tit, *Cyanistes caeruleus*) and olive-green (the great tit, *Parus major*).

Tits—in fact, all birds—cannot digest milk. Unlike mammals, they lack the enzymes needed to break down the lactose. But the layer of cream that collects at the top of old-fashioned, unhomogenized milk contains very little lactose, and a hungry bird in winter could do worse than supplement its fat intake with a bit of rich cream snatched from the neck of a milk bottle. And that is exactly what tits had been doing for a while in the late nineteenth and early twentieth century in England and elsewhere in Europe, when milkmen were still in the habit of leaving open bottles of milk on people's doorsteps in the morning. Before the resident mammal would have time to open the door and bring the bottles into safety, a tit would swoop in, land on the neck of a bottle and dip its beak in the cream inside, consuming up to an inch of the coveted food.

Unfortunately, the very first stages of the ensuing game of attrition between human and bird are lost in the mists of time. Presumably, it was a matter of racing to the front door as soon as the milkman's cart was spotted, not to give the tits a chance to steal any cream. Tits, not be outdone, would be hanging out near people's doorsteps at milk delivery time to still try to get there first. In any case, at some point in the early twentieth century, milk suppliers began closing their bottles with wax-board tops. This only gave momentary respite because in 1921, in Southampton, tits began prizing these off, or stripping away the cardboard layer by layer until the cap was thin enough to be pierced by the bird's sharp beak. Changing the cardboard caps to aluminum ones did not help for long either: by 1930, tits in ten different towns all over England had learned how to open the metal-topped bottles. When faced with a metal cap, they would hammer a hole in and then pull away the foil in strips. They might also pull off the entire cap and fly away, holding it in one claw and, in a secluded spot, peck at the cream sticking on the inside of the cap. Under the birds' favorite tree, the cleaned and discarded caps would gradually pile up to a respectable refuse heap. But sometimes their greed was their downfall. More than once, say Robert Hinde and James Fisher, two English ornithologists who studied the behavior, blue tits would be found "drowned head first in the bottle, presumably because they tried to drink too deep and lost their balance."

Hinde and Fisher learned all this because, in a citizen science project *avant la lettre*, in 1947 they mailed hundreds of questionnaires to bird-watchers, naturalists, milkmen and milk-consuming homeowners, but also to doctors and other "people with a scientific training." Using the replies they received, they were able to piece together a detailed history of the epidemic-like spread of milk bottle attack skills among tits, *and* the human counter-measures, all over the British Isles and, in a Europe-wide follow-up questionnaire, also on the Continent.

In an article in *British Birds*, they published snippets of some of the responses they got, and these betray the depth of the frustration felt by humans in this battle of wits with their mouse-sized opponents. People were exasperated at how quickly the tits were at their milk bottles, often within minutes of the milkman placing them there. As if the birds were waiting for it! (They probably were, since one milkman complained that some tits did not even wait for him to deliver the bottles to a home, but rather raided his cart while he was out placing bottles on somebody's doorstep. And then as he ran back to his cart, other tits would befall on the bottles just delivered.) In one particularly wholesale onslaught, a gang of tits managed to open 57 out of 300 bottles left at a school before the schoolmaster had a chance to chase them off. In some areas, people would provide their milkmen with heavy metal lids, rocks, or cloths to leave on top of the bottles, but invariably the tits would learn how to remove those as well.

The maps that Hinde and Fisher published in their article show how the birds' bottle-opening skills spread. Intriguingly, they did not spread out gradually from the source of the invention in Southampton. Rather, tits attacking bottles seemed to pop up independently in many towns and cities and then from there the behavior would catch on locally. Individual tits rarely move more than six to twelve miles in a year, and yet, new towns more than twelve miles away from the nearest affected town would suddenly also be afflicted by cream-hungry tits. So, it is more likely that the behavior was invented independently by multiple, particularly clever birds that then were imitated by others. For example, in the Welsh town of Llanelly, hundreds of miles away from the nearest knowledgeable tit, only one house in a neighborhood of 300 houses suffered from a single thieving tit in 1939. Seven years later, all the tits in this neighborhood were doing it. And in Amsterdam, Niko Tinbergen saw great tits opening milk bottles both before and after the Second World War, even though during the war, and in the lean years immediately after, milk was not delivered and no pre-war tits would have survived to 1947, when milkmen gradually began doing their rounds again.

Over the past few decades, the tits seem to have finally been defeated by their milk-guzzling human adversaries. First, skimmed and homogenized milk, which lack the cream on top, became more popular. For a while, the tits managed to circumvent these by learning the cap color denoting old-fashioned fatty milk. But since then, aluminum-capped glass milk bottles have slowly been replaced by other containers, and the traveling milkmen themselves have all but disappeared, in favor of the supermarket. Today, very few homeowners still know the infuriating feeling of finding their milk vandalized by neighborhood birds.

This ongoing battle between birds and bottles keeps inspiring urban biologists, because many mysteries remain. How did the bottle opening skills pass from bird to bird? Are city birds perhaps better or quicker at learning such new tricks or acquiring new tastes than rural birds? And, if so, why?

The answer to that first question—how a new trick learned by one clever bird is then passed on to others—was recently revealed by an Australian researcher, Lucy Aplin, working at Oxford University. Aplin's research takes place in Wytham Woods, that same forest near Oxford where Bernard Kettlewell lived in a trailer and collected peppered moths in Chapter 8. But these days, researchers use much more snazzy equipment than Kettlewell's muslin gauze sleeves. Aplin set up automated computerized "puzzle boxes" all over the forest. A puzzle box is the biologist's devious way to assess an animal's problem-solving skills. It usually is a contraption that requires a certain set of actions before it releases a reward in the form of a tasty morsel of food. In the case of Aplin's tits, the puzzle box consisted of a plastic chest with a stick to perch on. Right by that perch was a door that a bird could slide either to the right or to the left by pushing it with its beak. Behind it, it would find a dish with delicious live mealworms.

And that was not all. Being next door to a dense cluster of restless biologists, the great tits of Wytham Woods are an intensively scrutinized group of birds. Each, for example, has been fitted with a minuscule transponder chip in a leg ring. By placing antennae at nest boxes and feeding tables, researchers can keep

track of the personal history of each bird: how old it is, with whom
it had built a nest, but also: who are its friends, with which other
tits does it like to hang out? These individual identification codes
were picked up by an antenna hidden in the perch on Aplin's
puzzle box each time a bird landed. Switches in the plastic door
would register if the bird managed to open it, and—a crucial point,
as we shall see—it detected the manner employed: by pushing it
left or by pushing it right.

Wytham Woods, at least as far as its great tits are concerned,
is divided into eight sections, each harboring about a hundred
tits, which interact more with one another than with birds from
the other sections. The Oxford tit-watchers call those sections
"sub-populations." In each of five sub-populations, Aplin caught
two male birds that were given the honor of being the puzzle
boxes' early adopters: she taught these ten birds how to open the
puzzle box by letting them watch captive birds that already knew
the trick. Some she trained to open the door to the right, others
to the left, and she made sure that the two birds from the same
sub-population learned the same version of the puzzle (either
both pushing left or both pushing right). The enlightened birds
were then released back into their native sub-population to pros-
elytize the puzzle-box gospel, while Aplin set up a battery of
puzzle boxes, stocked with mealworms, all over the forest.

For four weeks, the circuits of switches, antennas, and digital
hardware in the puzzle boxes whirred away, constantly recording
the comings and goings of birds and the left- or rightward sliding
of the door. When the feast was over, Aplin packed up her puzzle boxes, downloaded all the accumulated data and began ana-
lyzing. She discovered that the majority of the tits, in the five
sub-populations where she had released the puzzle-box-savvy
birds, had learned how to open the door. But in the sub-populations

without any "trainer," only very few birds figured out how to deal with the boxes—less than 10 percent in one case.

It was also clear that this knowledge passed through the sub-population via networks of friends: the best pals of the ones who had been trained were the first to pick up and then pass on the new knowledge. Since the set-up recorded the exact moment at which each individual bird learned the trick, Aplin could actually watch the meme spread throughout the birds' social network until nearly everyone knew about it. And this is where the two alternative right-pushing and left-pushing door-opening solutions came in handy: in each sub-population, a tradition of door-opening took hold that depended on how the original trainers had been trained. If they had learned to slide the door to the right, that was how all tits in their sub-population eventually did it, and vice versa. Even a year later, Aplin found, this local box-opening custom was still there.

What the great tits of Britain show is that some animals can learn how to crack human code and then let their best friends in on the secret—at least until humans come up with a countermeasure. That's how humans and city-dwelling animals are constantly at loggerheads, but for such information to be learned and passed on among animals, certain faculties are essential. First, the animals need to have a kind of problem-solving intelligence; the kind that helped blue and great tits understand that breaching an aluminum top on a bottle would give access to delightful cream underneath. Secondly, they need to have neophilia—an attraction to unknown objects; when the first glass milk bottles arrived, some tits were not freaked out by them, but instead began exploring them for possible nutritional benefits. And, finally, they need to be tolerant of angry milkmen, tea-cloth-wielding homeowners, and close proximity of people in general.

Clearly, the tits that successfully attacked milk bottles, or Lucy Aplin's puzzle boxes, benefited from the fact that they were tolerant, problem-solving neophiles. But this is not always the case. Under more natural circumstances, it is often safer to be shy, conservative, and neophobic. In an environment that has been stable for a long time, humans and other large animals are better avoided since they can be dangerous—objects made by humans tend to have lethal moving parts, so better be safe than sorry.

But in cities, such traditional behavioral adagios might need to be reconsidered. Humans bring in their wake a superabundance of food, they create shelter and nesting sites, and generally offer new opportunities. Moreover, at least in cities, humans tend to be favorably disposed toward most small birds and mammals and not likely to harm them (though their pets might). Finally, humans are forever creating new stuff. Sometimes, like the McFlurry ice-cream cups hedgehogs get their spiny heads stuck into, these new objects are dangerous, but often (think milk bottles) the advantages outweigh the hazards. In other words, we might expect that city animals evolve to become better at exploiting their human neighbors. Not because some gene for opening bottle tops spreads in the population (surely no such gene exists), but because genetic tendencies to be tolerant and more inquisitive (and such genes *do* exist) will help an animal to quickly learn how to take advantage of humans and their ever changing ways. By enabling quicker learning, such genes will spread—and the species will evolve in the city to be a more street-smart version of its former rural, stuffy self.

There's actually evidence that this is the case—that city animals are fearless problem-solvers with a penchant for anything new. Some of this evidence comes from the island-state of Barbados, where McGill University in Montréal, Canada, owns a field center. Located on the fringe of the city of Bridgetown, generations of McGill staff and students have done their field teaching and graduate research projects there. The field center has a perfectly fine canteen, but hey! it's the sandy, sunny Caribbean coast, and the lavish Colony Club is right next door—so quite a few hours of pre- and after-fieldwork lounging are actually spent there. It was at the Colony Club's immaculately decked-out tables that in the year 2000, a few McGill biologists first noticed how cheeky Barbados bullfinches (*Loxigilla barbadensis*) deftly opened the paper sachets of sugar that had been intended for human consumption. Pretty much like the British blue tits on milk bottles, a bullfinch would hold a sachet with one claw and use its heavy beak to pierce the paper wrapper and gobble up a few mouthfuls of sugar before flying off. Later, the bullfinches were seen mastering other restaurant table manners, like opening sugar bowls (by turning over the heavy ceramic lids with their beaks) or stealing coffee creamer. "When you sit at a terrace in Barbados, it's almost guaranteed that you will share your table with bullfinches," says graduate student Jean-Nicolas Audet.

For Audet and post-doc Simon Ducatez, studying these bullfinch behaviors provided the badly needed justification for spending extended periods at the restaurant tables of the Colony Club. Eventually they even managed to persuade their supervisor to let them conduct a part of their "field" work at the Colony Club. And also at the nearby Coral Reef Club. And the sumptuous Royal Pavilion. But Barbados is not all city and coastal resorts. Although densely populated (on average nearly 700 people per square kilometer) and heavily urbanized, the northeast corner of the island is still rural. So Audet figured it might be interesting to see if the problem-solving skills of rural bullfinches matched those of the urban ones or not.

To study this, he devised two types of puzzle box. Both were made of transparent plastic and contained seeds as a reward, but one (the "drawer" box) could be opened either by tugging on a drawer or by pulling off a lid, while the other type (the "tunnel" box) required both actions: first tugging, then pulling. Audet caught twenty-six urban bullfinches and twenty-seven rural ones and, in the field center, tested if (and, if so, how quickly) they figured out the puzzle boxes. As it turned out, all birds managed to do the drawer box, but the urban birds were twice as quick at it than the rural ones. The more complicated tunnel box was solved by only thirteen of the urban bullfinches—but the rural ones did even worse: only seven of those figured it out, and it took them on average nearly three times as long as the city birds. Clearly, urban bullfinches will be better at coming up with new ways of accessing human-provided food. Whether the urban finches actually carry different genes for problem-solving skills from the rural ones is up for debate. Audet suspects the island might be too small for that and the birds too peripatetic. Then again, if the benefits are great enough, natural selection can still go against the stream, as it were, and slowly build up genetic differences.

Problem-solving was key trait number 1. But for an animal to even approach a problem to be solved, it needs to be less than circumspect about new and unfamiliar objects in its environment. Preferably, it needs to be neophilic: keen to approach and investigate anything that is out of the ordinary. In other words, it needs to be curious.

Over the years, experimental biologists have had a field day (literally) devising experiments to test urban animals' neophilia. What is more fun than cobbling together weird and wonderful objects that don't look like anything and confronting unsuspecting experimental animals with them? It's like candid camera for biologists. In the name of urban behavioral biology, common mynahs in Australia have been exposed to green hair brushes and yellow rolls of tape, English crows to works of modern art pieced together from chip bags, jam jars, and polystyrene fast-food containers, and chickadees in Tennessee to wonderful towers made from Duplo blocks. And in virtually all cases, urban birds approached these weird objects quicker and with more interest than wary rural birds.

One particularly thorough study is worth mentioning here. Piotr Tryjanowski and colleagues studied 160 bird feeders in and around Polish cities. Half of the feeders they adorned with an object "brightly green and made out of gum with a tuft of hair." "We never observed anything even vaguely similar in the field making it highly unlikely that local birds responded to this object as anything but a novel object," they write in their *Scientific Reports* article. Other feeders were left unmodified. Lo and behold, for the four bird species that visited the feeders most (great tit, blue tit, greenfinch, and tree sparrow), the rural visitors were neophobic: they avoided bird feeders with the weird green thing on the roof, whereas in the cities, the reverse took place: here, birds actually flocked to the pimped-up feeding tables.

After problem-solving and neophilia, the third and final personality trait that the urban environment will select for is tolerance: a reduced fear of humans. In a 2016 article in *Frontiers in Ecology and Evolution*, a team led by Matthew Symonds of Deakin University in Australia compared forty-two different bird species for their so-called FID: flight initiation distance, the average distance a human needs to approach a bird before it takes off.

They found that, across all these bird species, the urban version was more tolerant than the one from the countryside. Not only that, but the difference was greater the longer the birds had been residing in cities. For example, jackdaws (*Corvus monedula*) in cities (which they had already colonized by the 1880s) only get spooked if humans come closer than 8 yards, whereas in rural areas they will take off at 30 yards distance. Great

spotted woodpeckers (*Dendrocopos major*), on the other hand, urbanized only since the 1970s, still have similar FIDs in city and
countryside: 8 and 12 yards, respectively.

That positive relation with the time since first city-dwelling is
important, because it shows that tolerance has actually *evolved*. It
is unlikely that over the generations, wariness would wane by each
bird generation *learning* to be a tad less careful around people than
their parents were—you would expect this to happen more quickly.
Rather, if there is a benefit to being more tolerant, *genes* for tolerance can gradually accumulate and the species' personality will
evolve. Such an explanation is particularly likely, since the same
researchers also found that a bird's tolerance has nothing to do
with the size of their brains: brainy birds did not become tolerant
of humans more quickly than, well, bird-brained birds.

It seems likely that problem-solving, neophilia, and tolerance
all are prone to urban evolution, and we will actually see examples
of this later, when I will present you with some of urban evolution's *pièces de résistance*. For the time being, let's remember that an
important aspect of the evolutionary pressure urban animals are
under is their continual arms race with urban humans over gaining access to their food and other resources.

The evolutionary landscape of the city is now nearly completely
revealed to us. There are close encounters of the first kind—the
tough but static physical and chemical structure of the city (heat,
light, pollution, impenetrable surfaces and all the other urban
features we saw in Section II of this book). Evolution as a result of
such encounters may come to a standstill when the perfect adaptation is reached. Then there are the even more exciting close encounters of the second kind. These happen where urban animals
and plants interact with aspects of the city that are not static,
namely where they involve *other* animals and plants, including
humans—all of which could, in principle, respond by changing

themselves. This kind of encounter is all the more exciting because it may lead to "Red Queen" evolution: evolutionary arms races where both partners keep finding new ways to gain the upper hand. In theory, such evolution never stops.

Yet there is one final part of this urban evolutionary landscape that we have so far skirted around. In the previous chapters, we have seen close encounters of the second kind involving interactions *between* species. But what about that particularly close encounter *within* a species? Males and females of the same species also evolve to adapt to each other—we call this sexual selection. It would be naïve to think that there is no urban impact on the amorous animal.

16

SONGS OF THE CITY

EVERY YEAR IN SEPTEMBER, I ORGA-
nize a general orientation course for the new Leiden
University biology students entering our master's program in
Evolutionary Biology. In the first week, we always do some
urban ecology and evolution. We go searching for grove snails
using a smartphone app that records whether snail shells are
more brightly colored in the urban heat island (I'll say more
about that toward the end of this book). And we do an unusual
field practical. Most of the students view the entry in the day's
program with a puzzled frown. "Erm . . . What is 'urban . . .
acoustic . . . ecology'?" they ask. Just wait and see. We gather
outside the biology building waiting for the afternoon's instruc-

tor, my colleague Hans Slabbekoorn, who is—you guessed it—
an urban acoustic ecologist.

At 1:30 p.m., he emerges. Khaki shirt and shorts, with long
graying hair and, somewhat uncharacteristically (for an acoustic
ecologist), a pair of binoculars around his neck. He hoists his
bag, adorned with a Pacific Northwest native pattern, onto his
shoulder, stands himself in front of the expectant group of thirty
or so students, and outlines the afternoon's activity. We humans
are very visually oriented animals, he explains. We view our surroundings first and foremost with our eyes. But for a biologist, it
is also very important to be aware of the acoustic landscape, as
many animals communicate by sound. As an exercise to stimu-
late that auditory awareness, Slabbekoorn explains, "We're going
on a silent walk. We will walk in a line, not talking, in total still-
ness, to become more aware of all the sounds around us." It would
be even better, he adds, to also close our eyes, but that would pose
too many navigational challenges.

So, off he heads, through the residential area near the university, and then into a nearby urban park. The students have little
option but to follow him, and I bring up the rear. After some
initial giggles and shooshings, the group of students morphs into
a taciturn chain gang walking along the main road. Cars appearing from side-streets stop for us with revving engines, and pedestrians stand still to look at this strange long file of people walking
in complete silence along a busy thoroughfare. Some passersby
make duck-quacking noises in ridicule. But we manage to maintain our silence and do as Slabbekoorn wanted us to do: take in
the soundscape of the city.

And it works. We hear things we would not have heard otherwise. The different types of growl of diesel- versus petrol-powered
vehicles, the grating, clangy sounds of old battered bicycles passing by, jetliners flying overhead, the incessant crashes from a

nearby building that is being demolished . . . But also the wind rustling in the reeds, the leaves of poplar trees rattling, a robin doing its little waterfall-like song, a hammering woodpecker, the bubbling sounds of a nuthatch, the screams of ring-necked parakeets flying over . . . Subtle details, too: our footsteps changing in character when we move from pavement to shell-lined paths in the park, and a grasshopper's papery song falling silent as we pass by.

In the end, we gather on an open field, surrounded by tall trees, with the old university hospital dorm being torn down in the background. Slabbekoorn: "This used to be one of the quietest places in Leiden. The tall hostel shielded the din from the busy road behind it, and we'd be far away from the urban center." These days, the area has become noisier: the dorm building is being demolished and on top of that, the city sounds now penetrate the park unhindered. Upon prompting by Slabbekoorn, the students reveal how much they picked up. One student remarks how the traffic actually seemed to become louder as we entered the forest. "That is temperature inversion," Slabbekoorn explains. On the forest floor it is cooler than out on the street, and the traffic sound becomes trapped in the layer of cool air at ear level.

"Let's close our eyes for a second," Slabbekoorn says, "and listen to the urban sounds." Of course, we initially hear only the fits and starts of the heavy machinery gnawing away at the carcass of the student dorm, and the staccato plops of a heavy motorcycle passing by, but Slabbekoorn asks us to ignore those and focus instead on the low-level background sounds of the city. And indeed, when we train our ears to it, in between all those discrete urban sounds, there is a nearly imperceptible carpet of constant, low-pitched rumbles, waxing and waning irregularly. It is the city's breath: a cacophony composed from the combined sound waves emanating from the engines, brakes, and horns of countless motorcycles and automobiles, the steel on steel of passing trains, the jet engines of airplanes, air-conditioning compressors and other machinery, construction pilings, voices and shouts, music from loudspeakers, and so on, all mixed into the grayish porridge that we call noise, muffled and channeled via the labyrinth of buildings and streets. In Europe, 65 percent of the human population is exposed to urban background noise louder than that of constant rainfall. And the animals in the city that try to make themselves heard have to contend with all this as well.

Dealing with background noise is not unheard of, of course. Natural habitats can be loud, too: frogs that live near streams or waterfalls, or birds in rocky canyons where every sound is amplified by echoing, know this problem all too well. Or think of a cricket trying to make itself heard by another cricket in a tropical jungle full of animals' yells, whoops, buzzes, and whirrs. Some of the solutions from those pristine environments are surprisingly similar to the ones that urban animals come up with, Slabbekoorn explains, and he points to the wheezy, high-pitched "bicycle-pump" melody of a male great tit singing from a poplar tree behind us, "dee-du, dee-du, dee-du!"—crystal clear despite the backdrop of urban rumble.

It is the great tit's "dee-du," and all the variations on that theme, uttered to attract females and to repel other males, that brought Slabbekoorn his early fame, when, in spring 2002, he and his student Margriet Peet began recording *Parus major* all over Leiden. From April to July of that year, they became a familiar sight to the city's residents, hauling recording equipment, a directional microphone, and an omnidirectional microphone on a five-yard-long pole from neighborhood to neighborhood, like a pair of traveling acrobats. At thirty-two places, ranging from quiet residential areas like the park where our acoustic ecology lab took place, to busy city center crossings or highway roadsides, they set up their microphones to record the songs of territory-defending male great tits (females don't sing) with the directional microphone, and pick up ambient urban rumble with the omni-directional one (from the tits' vantage point, hence the five-yard pole). And to average out the effects of the time of day, they paid each bird three visits: to record sounds before, during, and after the rush hour.

The results, which Slabbekoorn and Peet published in a very influential one-page article in *Nature* in 2003 (more than 700 other publications have cited it since), revealed the birds' struggles to make themselves heard above the din of traffic. Pitch played a crucial role in that. Most urban noise is concentrated in a low-frequency band of up to 3 kilohertz. The repertoire of the great tit spans a range from 2.5 to 7 kilohertz, the lowest notes overlapping with the urban noise. Slabbekoorn and Peet discovered that tits in noisy areas of Leiden deal with this by raising the pitch of their songs to above 3 kilohertz to avoid being drowned out by the sounds of the city, whereas those in quiet neighborhoods also use tones that go down to below 2.5 kilohertz.

As early as the 1970s, zoologists studying the famous great tits of Wytham Woods had discovered that the birds adjust their song to their surroundings: birds in open woodland sing higher-pitched songs than in forests because dense vegetation tends to muffle the higher notes too much. But Slabbekoorn was the first to discover that the birds apply the same musical strategy in our urban habitat. Since his ground-breaking studies, dozens of bird species in cities in many different countries have been seen (or rather, heard) to do the same: Chinese bulbuls (*Pycnonotus sinensis*), in Asia; song sparrows (*Melospiza melodia*), in North America; rufous-collared sparrows (*Zonotrichia capensis*), in South America; silvereyes (*Zosterops lateralis*), in Australia . . . All over the world, urban birds sing higher, and probably also louder, than the same species in a quiet rural setting. And not only birds: also the Australian Southern brown tree frog (*Litoria ewingii*) croaks higher in Melbourne than in the surrounding country-side, and grasshoppers of the species *Chorthippus biguttulus* along noisy German roads sing a shriller song than in quiet meadows.

While he is pleased that his work has spawned so much new research, Slabbekoorn says that many questions have surfaced along the way. Do song-determining *genes evolve* in the city because males with inaudible baritone genes can't woo any fe-males, while the tenors get all the girls? Or do they *learn* to drop the deeper-voiced songs from their repertoire? And, if they learn, do they do so by imitating their fathers or rival males, or by keep-ing track of which songs have the most impact? And what about plasticity? Could it be that animals growing up in a noisier place automatically grow squeakier voices? Slabbekoorn and his fellow urban acoustic ecologists are still grappling with these questions, but they seem to depend on the kind of animal.

One of Slabbekoorn's students, Machteld Verzijden, took the lab's sturdy microphones to the busy Rotterdam-Amsterdam A4 highway just outside of Leiden. Here, despite the din, lots of male chiffchaffs (*Phylloscopus collybita*, a sleek brownish-gray warbler) advertise their territories by singing their characteristic monoto-nous "chiff-chaff" song throughout the breeding season. Like in the great tits, her recordings revealed that the frequencies of every "chiff" or "chaff" were, for the birds singing close to the highway, about 0.25 kilohertz higher than for those along a quiet river a half mile away. But Verzijden did not stop there. She brought a boom box to her riverside site and, while the rural chiffchaffs were singing, gave them a taste of the noise level that their hard-shoulder brethren had to cope with by playing them loud traffic noise at close range. The results? An individual chiffchaff will immediately raise the pitch of his song if there is a lot of noise around: her rural chiffchaffs also increased their chiffs and chaffs by about 0.25 kilohertz as soon as she flipped the boom box switch.

Clearly, no evolution there: the deep-voiced riverside and high-voiced highway chiffchaffs do not differ genetically; they simply adjust their song to the ambient noise. But in some other animals, it is not that simple. The songs of frogs and of many non-songbirds like flycatchers and doves, for example, are much more stereotyped. They are hard-wired from birth and cannot easily be changed simply because humans are creating a racket. And the same goes for songbird *calls* (short utterances for expressing alarm or keeping in touch with one another). And yet, the songs of frogs and non-songbirds, as well as the calls of songbirds, are all higher-pitched in cities—while they are less likely to be simply adjusted by the animal itself.

The results from the Department of Evolutionary Biology at the University of Bielefeld, where they study those *Chorthippus biguttulus* grasshoppers in *Autobahn* roadside verges, are even cuter. When PhD student Ulrike Lampe took immature males (which do not yet sing) from the bustling roadsides and from the sleepy countryside to her lab, put them in separate boxes and let them grow until they were mature and ready to sing, the songs of the ones that had been brought from the roadsides were about 0.35 kilohertz higher. This would seem like insurmountable proof of urban evolution, since these insects had never had the chance to learn about urban noise and yet as soon as they hatched into adult grasshoppers, they began singing at the perfect pitch. But even here, the reality is probably a bit more complicated. For when Lampe split the immature insects into two groups, and let one group grow up in a quiet lab, and one in a lab where she played constant recordings of traffic noise, the ones from the noisy labs grew up to sing at a slightly higher pitch than the ones from quiet labs—regardless of whether their original habitat had been roadside or meadow. In other words, the grasshoppers' urban acoustics are a bit of both: part evolution (nature), part plasticity (nurture).

Now this chapter is about sex, so talking about the urban acoustics of only the sex that sends a sexy signal is only half the story. In fact, it is no story at all, if we do not take into consideration how this affects the receivers for whom all these love songs are intended.

For a male great tit singing in his territory, those receivers come in multiple kinds. First, there are the neighboring rival males, always eager to step on your turf or have illicit affairs with your female. Secondly, there are the females. Ah, the female *Parus major*! She needs to be wooed into building a nest with you. Then, every day she needs to be persuaded to have her daily egg inseminated by you and not by one of the neighboring males; and then there are all the neighboring females which you might entice into having a quickie with you. This whole theater of great tit socio-sexual opportunities, threats, decisions, and interactions is played out during the dawn chorus, when territorial males are flitting about nervously, loudly broadcasting their "dee-du" calls, and meanwhile keeping an eye on their own females and on rival males, as well as ogling female neighbors.

What would be the effect on this theater of sexual rivalry if the medium that plays a central role is compromised by urban noise? This was the question that Hans Slabbekoorn and his colleagues were bouncing off one another around ten years ago. And, as so often with pressing academic questions, it did not take long before two PhD students were saddled with them. The first, Millie Mockford, then at the University of Aberystwyth, focused on the rival males. In twenty cities across the UK, Mockford placed a loudspeaker in the territory of an urban tit and played him a (low-frequency) song recorded outside that same city and also a high-frequency urban song. Then, she observed how the male reacted to this artificial rival. And she also did the opposite experiment: playing urban and rural songs in a rural male's territory. What she saw through her binoculars was that the males would become much more agitated by a song that matched their own habitat. In other words, an urban tit was more offended by an urban song than by a rural song, and vice versa.

The other study, which had a female focus, was the work of Slabbekoorn's student Wouter Halfwerk. He discovered that urban tits face a grueling quandary. Acting pretty much like the great tit secret service, Halfwerk closely watched a population of great tits breeding in thirty nest boxes in the Netherlands. By checking these nest boxes regularly he knew exactly when the females were fertile and when they laid their eggs. DNA tests told him which chicks were fathered by the male whose territory the nest box was in. If that weren't enough of a breach of their privacy, he also wire-tapped their homes, placing one microphone inside the nest box and another one outside. This way, he could record the male songs and also his female's soft-voiced encouraging responses, as well as the tell-tale scratching and wing-flapping when the female left the nest box, ready for her early-morning copulation.

What this surveillance operation told Halfwerk was that females just *luuurve* a deep-voiced male. The lower a male's calls, the more likely his female was to seek his company at the moment she was expecting a new egg. Sounds romantic, but the flipside of this is that females whose male did not achieve a deep sexy song would regularly sneak out of the nest box before dawn to seek another male's attentions. Sure enough, DNA tests showed that the high-singing males were being cuckolded: one or more of the chicks they were raising were fathered by the neighbors.

Now, Halfwerk's nest boxes were all in a quiet forest. To see what the effect of urban noise was, he had to add it manually. So—another secret service tactic—he subjected them to continuous noise until the tits gave up their secrets. On top of the nest boxes, he installed loudspeakers that were connected with mp3-players and pumped constant traffic noise into the poor birds' nest boxes. Then he broadcast pre-recorded high-pitched and low-pitched male songs from a loudspeaker placed outside the box. Only when the songs were high enough to be heard over the traffic raging inside the nest box would the female emerge, expecting to be mounted by the male (but then wasn't, as there was only Wouter Halfwerk and his loudspeaker).

What these two studies show is that great tit sexual evolution might be on diverging tracks inside and outside the city. Their songs, the level of monogamy, and what entices a response in both males and females might all be drifting away from what is the norm outside the city. And probably similar things are happening in other urban songbirds whose voices have been cranked up a few notes.

During this long exposé by Slabbekoorn, some of the students have decided to stretch out on the grass, while the ones still standing begin to fidget. Clearly, as far as they're concerned, there's only so much you can absorb about great tits, loudspeakers, song frequencies, even about early-morning copulation, and maybe it is time to consider this urban ecology practical finished. Slabbekoorn takes the hint and begins to head back to the university. But at the edge of the ditch between the poplar-lined road and the biology department, he stops for one last twist.

"It's not just about pitch, you know," he says. "Urban noise can have an impact on bird acoustics in many different ways."

Australian silvereyes, for example, not only sing higher but their songs are also shorter and the musical elements are more spaced out, possibly to allow the song's echoes against buildings to peter out—the same reason why an orator in a large stadium will speak more slowly, so as not to interfere with his own returning echoes. Robins in noisy parts of the city of Sheffield (and, one assumes, in other cities, too) sing more during the night, when it is quieter. And then there are the planes in Spain's plains. In the flood plains of the Jarama, the dawn chorus of songbirds starts earlier where the river runs alongside Madrid airport's runways. To stay ahead of the roar of the day's first incoming and outgoing air traffic, the local blackcaps, warblers, cuckoos, and finches advance their internal alarm clocks by up to forty-five minutes.

Yet there are some urban animal-sound interactions where adaptation is not possible, says Slabbekoorn. He points at the ditch behind him. "According to Dutch legislation, a building project must be shifted if a ditch would need to be drained that contains the weather loach, a protected fish species. But the piling *next* to that ditch will kill just as well: the transduction of sound in water is very good and also the transfer into the watery body of the fish. The piling noise will rupture fish ears or their swim bladder." With that, we turn around and set ourselves in motion again, silent once more, but for different reasons this time.

17
SEX AND THE CITY

IN A SUBURBAN SAN DIEGO WALKWAY
stands a red women's bicycle with a rusty chain, parked among
a few other bikes and some gardening tools. On its rear carrier, a
plastic white-and-blue child's seat, inside which lies, upturned, a
Styrofoam cycling helmet. A mother returned home on a Friday
afternoon after picking up her child from school, parked the bike,
and then helped her child out of the seat, who immediately
wanted to run off to play in the yard. "Hey, what about your
helmet, sweetie?" she would call after her, then help unbuckle
the impatient child's helmet, and cast the thing in the back of the
plastic seat. But next Monday morning's rushed commute to
school was stopped in its tracks because of an unexpected turn of

events: "Well, what do you know, sweetie? A birdie has made her nest in your helmet!"

This is what I imagine may have preceded the taking of the photograph printed on page 189 of the April 2006 issue of the journal *Trends in Ecology and Evolution*. The reason that such a family snapshot appeared in the pages of this esteemed scientific journal is that the bird who took up residence in this San Diego shed is not just any bird. It is the dark-eyed junco (*Junco hyemalis*), which, in that part of North America, would normally nest exclusively in the conifer forests high up in the mountains. Until 1983, its breeding range lay hundreds of miles away from San Diego, at elevations of 1,600 to 3,200 yards. But in that year, to the surprise of local birdwatchers, dark-eyed juncos took up nesting in the coastal, urban environment of the University of California campus. Those first colonists were presumably a few mountain birds who had spent the winter near the coast, but, unlike all such winter visitors before them, decided not to migrate back up in spring. Instead, they stayed put and began nesting in the ornamental shrubbery among the campus buildings; and, yes, eventually also in bicycle helmets. Over the ensuing years, the colony grew steadily and they had reached a size of some 160 birds by 1998, the year in which biologist Pamela Yeh started studying them for her PhD.

The birds are not much to look at: sparrow-sized, drab brown and slaty gray, with a bit of white in the outer tail feathers. But it was those white tail feathers that interested Yeh, because these play an important role in the birds' love life. When a male junco fancies a female, he will try to impress her by hopping about, drooping his wings and fanning out his tail to display those bright white flags. In the 1990s, researchers studying the birds in their native range proved the effectiveness of this display with a simple cut-and-paste experiment. They cut off males' tail feathers and enhanced them or toned them down by super-gluing feathers that had either more or less white on them than the original ones. They discovered that females consistently went for the males with the whitest tails. Apparently, a male with a whiter tail, whether this is his natural color or not, makes a female junco's heart throb.

But why would this be so? What benefit could a female possibly gain by choosing a mate with a bit more white in a few of his feathers? Or, for that matter, with a slightly lower-pitched song, as we saw in the great tits of the previous chapter? For the answers to this intriguing question, we have to delve a bit deeper into the field of sexual selection. We'll get back to Pamela Yeh and her urban juncos after that.

After natural selection (where the environment selects), sexual selection (where the opposite sex does the selecting) is the second great force of evolution. Any genetic property that will make an organism sexier, and thereby gain more or better sexual partners, will see a higher representation in the next generation. As we have learned, such a change in genetic representation is, by definition, evolution. So sexual selection, as well as natural selection, will make a species evolve.

As an example, think of the male sage grouse (*Centrocercus urophasianus*) with his bizarre starburst tail, white ruff, chest sacks of bare yellow skin, and tiara of plumes on his head. Over thousands of years, female sage grouse (which are plain, stripped-down versions of the male) have been most impressed by males with the greatest starbursts, whitest ruffs, most in-your-face throat sacks, and longest head plumes. Those were the ones that got to sire the females' offspring, leaving all the less striking males without any descendants and their less-sexy genes as genetic dead-ends.

But there is another way in which sexual selection can work. Not by active choosing by the opposite sex, but by battling among sexual rivals; the winner then takes all. Imagine male rhinoceros beetles with horns that are so big that they can topple over all their competitors and always get to mate with any nearby female. These males will pass on their big-horn genes, at the expense of the less-endowed males, and the average horn size of the beetle species will increase over time—at least until horns get so big that they become a liability (and then natural selection kicks in by removing those ridiculously-large-horn genes).

These two examples are about sexual selection doing its thing on males. But in principle, sexual selection can work in both directions. Both females and males will select partners that seem best at producing large and high-quality litters. However, in practice, the two sexes differ in how much they stand to gain by selecting a "good" mate. In many species, the female invests a lot of time and energy in nurturing only a handful of offspring. For her, choosing the best father with the finest sperm is of the utmost importance—one wrong choice and her babies would be saddled with inferior genes. Males, on the other hand, often do not invest as much as females. For many a male, choosing the wrong female may come at a cost no greater than an ejaculate and a few wasted minutes. As a result, the evolutionary premium tends to be greatest on a female choosing the right male, and less so the other way around.

But how do you pick a good male? First of all, the answer to that profound question depends very much on what is relevant in your particular species' ecology. In some species, it's important to get a male that is good at defending a territory, in other species one that is good at finding food for mother and young. And in yet another species, males do neither, and all you need from him is his sperm. Still, knowing what you need does not entirely solve the problem, because how can you tell whether a male will be a good fighter, provider, carer, or sperm-donor without actually going so far as giving him a try? What a female needs is an "honest signal"; some sort of "flag" that can be used as a proxy for a male's true quality.

This brings us back to the dark-eyed junco's tail. Those stylish white rims on the male's tailcoat are not just aesthetically attractive to females—they stand for much more. As one of Yeh's colleagues found out, males genetically predisposed to have more white in their tails also have higher testosterone levels and are better at fighting off competing males. It is not yet clear exactly how this correlation comes about, but what *is* clear is that females can thereby use male tail decoration as a handy shortcut to the most testosteroney ones. In the dark-eyed junco marriage market, that is important. Up in the California mountains, the breeding season is short. The brief period during which the insects that they feed to their young are plentiful is only long enough for one, maybe two broods. That means that claiming a territory with a bountiful supply of bugs and successfully defending it against encroachment by other juncos is vital. So, what a female junco wants is a burly male who can do that. And his tail feathers will give her an important clue.

But for the San Diego campus juncos, Yeh reckoned, life is very different. Instead of having to grapple with the limitations set by a cool montane forest, the Mediterranean climate on campus is so balmy that the birds can start nesting as early as February and, thanks to the irrigation system that negates the summer drought, keep this up throughout the summer and early autumn, raising up to four nests per year. The downside of campus life, however, is that you can be an easy target. The terrain is very open. The lawns, parking lots, and streets afford clear lines of sight for passing hawks, who regularly swoop down and snatch up a junco that ventures out into the open. So, with less need for a macho male, the urban females should evolve a taste for drabber males, while at the same time, the ones with most white on their tails would get eaten more by birds of prey. Yeh reasoned, with both these forces pushing in the same direction, the urban juncos should evolve to show less white on their tails. And, sure enough, that is exactly what she found. Compared with the mountain birds, the ones on campus had evolved about 20 percent less white. Back in 2002, that is. Has the trend toward less tail-white continued since then? Yeh: "That is a good question! We don't know. After an extended period away from this field site, we will be sampling it come 2018."

Something very similar seems to be going on in—wait for it—that favorite of urban biologists' birds, the great tit. Juan Carlos Senar of the Natural History Museum in Barcelona discovered that Barcelona's inner city tits have narrower neckties than countryside tits. Now you have to realize that for a male great tit, the source of all his power is his necktie. The width of that vertical stripe of deep black breast feathers is mostly genetically determined and is, like the dark-eyed junco's white tail flags, directly connected with his masculinity. Birds with broader ties are more dominant and aggressive, are better nest defenders and get to mate with better females. In other words, big tie tits are top tits.

So why would male tits' urban attire involve narrower neckties and thus be of the less macho kind? It could be, of course, that the city is the place of refuge that absorbs all the weaker males from the countryside that could not maintain a territory in the face of broad-tied bullies. But Senar's work shows that that is not the case. By keeping track of the leg rings of the 500 or so males of which he knew the tie-width, he could tell how well each category survived. As it turned out, in the countryside, the broader a tie, the greater the survival—as expected. In the city this situation was reversed: narrow-tied males fared well, and broad-tied individuals were dying by numbers. So the city, it seems, has its own rules for what makes a good male.

What the great tit and dark-eyed junco stories show is that the quality of a male in the countryside depends on different things from in the city. In these two examples, the brawler males, so able-bodied and highly-prized in the forest, for some reason seem to be less useful in the city. If so, it is to be expected that female tastes in males would evolve along with this, so that city females, too, would begin turning up their noses for those most macho of males. This would then lead to a change in the ornaments, the signals by which females gauge a male's quality so that the birds inside and outside of cities would actually begin to *look* different—perhaps to the extent that they would evolve into different species, something that we will encounter in the next chapter.

Lest the preceding pages gave you the impression that the city always promotes the evolution of "metrosexual" males, there are also cases where the opposite is true. Surely you remember from previous chapters that fragmentation is one of the hallmarks of animals' and plants' habitats in the city. This is the case for wooded areas (think of all those small parks in New York City, where white-footed mice are marooned in a sea of mouse-unfriendly cityscape), but also for watery places. Damselflies, for example, need ponds fringed by some vegetation where they can perch, patrol for prey, and lay their eggs below the water surface. Their larvae lead lives entirely submerged, even. So for damselflies to reach and colonize the ponds, ditches, and other minuscule water bodies that dot the city, they need to fly long distances.

Nedim Tüzün and Lin Op de Beeck, two PhD students at the University of Leuven, Belgium, figured that this need for being a strong flyer would show up in the urban ponds. They put their hunch to the test by catching nearly 600 males of the azure damselfly (*Coenagrion puella*), at water bodies in and around three Belgian cities, and then testing the flight endurance of each of those males in a flight tunnel. Their flight tunnel was a Plexiglas tube of two yards long and half a yard in diameter, closed at the top and placed at an angle. They would put a damselfly in a cup, release it at the bottom of the tube, and then let it fly up and away until it got tired and drifted back to the bottom of the tube. On average, the countryside damsels would tire out after three and a half minutes already, whereas the urban ones would stay aloft for more than twice as long. So, the students' expectation was borne out: the ones that had managed to colonize inner city ponds must have been the strongest flyers, and this left its signature in the flight performance of the urban damsels.

But what does this have to do with sex? Well, in these damselflies, the sexual game is played out primarily via competition between rival males. When they're in the mood for love, they zigzag across the water surface and pounce on any female that shows itself. The male who reaches her first will grab her by the neck with a pair of pinchers at the end of his body and then carry her to a secluded spot for a bit of mating—meanwhile defending her against other males. For each male that they had caught, Tüzün and Op de Beeck had recorded whether it had been caught *in flagrante delicto* or as a lonely, unmated male. When they then checked their flight endurance data, they found that, in the urban ponds, the males that had been found in the company of a female had a flight endurance some 40 seconds greater than the ones that were netted as bachelors. In other words, in these urban insects, natural selection first makes them evolve into stronger flyers, and this is then even further amplified by sexual selection. Not only do the strongest flyers colonize the city ponds first—they also are the first to get to the available females.

Let us now lean back and see what picture of sex in the city we have painted. First, like anywhere else, animals do their best to advertise themselves as desirable partners. In principle, the city's dating network employs the same channels for this as in the animals' original habitat: beautiful sounds, striking colors, impressive actions. But when we look closely, we see that the urban animal appreciates different things in a partner. So the sexual preferences evolve, as well as the qualities that are valued in the other sex. A desperate dark-eyed junco writing in the personals section of the *Campus Newsletter for Birds* would perhaps say, "Caring male with almost no white in tail seeks acquaintance with female to raise several nests in comfortable bicycle helmet. No fighter but excellent at catching lots of small mosquitoes"—something that the butch readers of the *Highland Junco Journal* might find laughable.

Secondly, the way these sexual signals are broadcast may change, because of interference with the city environment. Noise and light pollution could reduce or shift the bandwidth at which auditory or visual signals work. Or one kind of signal might even take over from another. For example, using a (rather hilarious-looking) remote-controlled "robo-squirrel," a team of researchers from Hampshire College in Massachusetts discovered that when urban gray squirrels (*Sciurus carolinensis*) alert each other of potential danger, they respond better to the sight of tail flicks than to the sound of alarm squeaks. In rural squirrels, they found, it is the other way around. Presumably, this is another result of urban noise, and you could imagine that the same shift from sound to sight may be happening in sexual messaging in some animals.

Or, for that matter, a shift from smell to sight. In their natural habitat, Indian gerbils (*Tatera indica*) encounter one another so rarely that they rely on scents left as chemical "flags" scattered in the environment. But in cities these animals live tightly packed, and they seem no longer to have a need for such long-distance olfactory communication. As a result, urban gerbils are losing their scent-marking glands.

Some interference by the urban environment with animal sex messaging may be even more insidious. Chemical pollutants with such exotic names as organochlorines, phthalates, alkylphenolic compounds, polychlorinated biphenyls, and polychlorinated dibenzodioxins are released into the environment from a variety of sources. They can be pesticides, additives to plastics, and industrial wastes. What these substances have in common is that they persist in the environment. Many have already been banned but since they can have a half-life of centuries or more, these pollutants will be a fixed feature of the urban chemical landscape for a long time to come. Another common property is that they interfere with sexuality. This is because chemically, they mimic certain sex hormones that animals use to fine-tune their sexual development, both physically and in terms of behavior. The result is the kind of sexual aberrations that we find, for example, in male alligators in lakes polluted with DDT, which have small penises and low testosterone. Conversely, female mosquitofish living in the run-off of a paper mill were found to have male body features and were hyper-aggressive and dominant. You can only guess at how evolution by sexual selection will be struggling to adapt to an interference of this magnitude—if that is at all possible.

Another type of anthropogenic insinuation into the delicate workings of an animal's sex life is called an "evolutionary trap." Unwittingly, we humans sometimes create stuff that latches on perfectly to the traditional ways in which a certain animal's courtship is conducted. Take the Australian satin bowerbird (*Ptilonorhynchus violaceus*), for example. Like all bowerbirds, the males of this species build amazing works of art to persuade females to mate with them. They create something akin to an ornamental garden, complete with passageways, entry lanes, and arrangements of beautifully shaped and colored objects that they harvest from their environment. Before that environment became domi-

nated by people, the satin bowerbird would use rocks, shells, flowers, butterfly wings, and beetle elytra for this purpose. But, these
days, humans supply an endless range of attractive artificial objects to add to his display. He is especially taken by bright blue
things—such as those rings that snap off the caps of milk bottles
when you open them.

Sadly, those rings turn out to be—quite literally—an evolutionary trap. Sometimes, when the male, excited about such a prize
find, is carrying it around in his beak, the ring flips backward
and becomes trapped behind the nape of the neck. Thus permanently muzzled, the male then strangles himself or slowly starves.
Our mindless interference with his pure aesthetic intentions
brought on his downfall.

Curiously, yet another Australian drinking habit wreaks havoc
on an animal's love life. In 1983, two Australian entomologists
published a short article in the *Journal of the Australian Entomological Society*, entitled, "Beetles on the bottle." Centerpiece of the
article were two photographs of a large, yellowish-brown jewel
beetle, *Julodimorpha bakewelli*, trying to copulate with a particular type of beer bottle, locally known as a "stubby." The animal
had mounted the round bottom of the bottle and was furiously
but unsuccessfully trying to penetrate the glass surface with its
long, brown penis. The stubby and its lover had been found on
the sand along a highway outside of the town of Dongara. When
the two authors searched the vicinity, they found several other
bottles with similarly infatuated male jewel beetles clinging on
to them.

It is not likely that the insects were attracted by any beer in the
bottle because, as the authors remind us, no Australian would ever
throw away a bottle that still has beer in it. Instead, the attraction
probably lay in the color and gloss of the glass, its curvature and,
crucially, its texture: a rim of small, evenly spaced tubercles that

had been pressed into the lower part of the bottle during the manufacturing process. This combination of features brought enough similarity with the backside of a female *Julodimorpha bakewelli* to make this discarded glassware irresistible to the male beetles. When the pair of entomologists laid out several new bottles in the area, male beetles did indeed descend on them within minutes.

Stubbies are an evolutionary trap not because they kill the beetles, but because they distract the males from their job of mating with real females. It is even possible that the males think they have found a particularly large, shiny, desirable female, by which all real females pale in comparison. If there are a lot of such superfemales (super-sexy, but, of course, also super-infertile) lying about, we would expect that the only hope for a way out of this evolutionary trap would be males that are genetically predisposed *not* to be turned on by beer bottles (maybe focusing on other female features, like scent). Only they will be able to reproduce and this could, eventually, cause the beetles and their sexual signals and preferences to evolve. It would not be the first time that a marriage was saved by weaning the male off his beer bottles.

18

TURDUS URBANICUS

THE GALÁPAGOS: A HANDFUL OF CIN-
ders tossed into the Pacific Ocean, where evolution has
cooked up unique ecosystems from the few ingredients of flora
and fauna that the South-American mainland has thrown it. A
world unto itself, with home-grown evolutionary trees of tor-
toises, giant and dwarf cacti, mockingbirds, iguanas, *Bulimulus*
snails, darkling beetles, and, of course, the islands' most famous
inhabitants: the fourteen species of Darwin's finches, each with
a beak shape to suit its particular walk of life.

They're not finches actually, but either tanagers or buntings.
(Ornithologists aren't quite sure.) They were only named "Darwin's
finches" in 1936, more than a century after the great naturalist

discovered them on his voyage on *HMS Beagle*. Nonetheless, Darwin's finches have become the poster children for evolution. Not only did Darwin push them as one of the prime examples of his theory—"One might really fancy that, from an original paucity of birds in this archipelago, one species had been taken and modified for different ends," he wrote—they have also taken center stage in the past forty-five years of cutting-edge evolutionary research.

Since the early 1970s, a whole dynasty of scientists, mostly working from the Charles Darwin Research Station on Santa Cruz, the archipelago's second largest island, have studied the birds and have revealed, in amazing detail, how Darwin's finches continue to evolve. The teams keep track of the birds' births and deaths, trysts and quarrels, food fads and nesting sites, and measure the sizes and shapes of their bills and bodies, year in, year out. They take blood samples, record songs, and run DNA tests. All this hard labor enables the biologists to watch, and even predict, the finches changing shape in real time. Each turn in the harshness of the climate, or the availability of a particular type of food, results in an evolutionary shift in the birds' physique. It's often a matter of fractions of millimeters, but it's measurable and real.

For example, on the island of Santa Cruz, the medium ground finch (*Geospiza fortis*) is in the process of splitting into two. You can tell by looking at their beaks. There are a lot of birds with small beaks, a lot with much larger beaks (up to almost twice as big), but not so many intermediate ones. Beak size translates directly to the kinds of seeds the animals can crack. Big-billed *G. fortis* finches have a bite that is more than three times as strong as small-billed *G. fortis*. So, they can handle the heavy-duty seeds like caltrop, whereas the weak-beaked ones are good at feeding on smaller and softer seeds of, say, grasses. The intermediate-billed birds, however, fall between two stools: their beaks are not strong enough to crack the large seeds and not small and delicate enough to efficiently manipulate the tiny seeds. So, during lean times, they will be more likely to starve, and natural selection grimly removes them. What's more, the beak size difference has sexual knock-on effects: large-beaked males sing a different song from small-beaked ones, large-beaked females prefer to mate with large-beaked males and small-beaked females with ditto males, so there is less genetic exchange across the beak-size divide. In other words, there is *speciation* going on: the evolution of two new, separate species where previously there was only one.

While the Darwin's finch has become the emblem of speciation in the wild, another bird has taken up that role for urban speciation: the blackbird, *Turdus merula*.

In 1828, the same year that Darwin befriended John Stevens Henslow, the Cambridge don who arranged for him to go on his *Beagle* cruise, a small book was published in Italy, entitled, *Specchio Comparativo delle Ornitologie di Roma e di Filadelfia*. The author was Charles Lucien Bonaparte. A wayward nephew of the better-known member of the Bonaparte family, Charles Lucien lived an incorrigibly zoological life. Having spent his youth in Rome and, after his marriage, the better part of the 1820s in Philadelphia, he published this "mirror" (*specchio*) of the avifauna of both cities.

His mirror consists of two columns. The birds of Rome are on the left, those of Philadelphia on the right, and everything is carefully arranged along the lines of the official bird classification of that day (in which Charles Lucien himself was one of the main authorities). On page 32, in the column for Rome, we find the following entry:

"69. TURDUS MERULA, L. *Merlo, Merla*. Comunissimo. Permanente; alcuni individui migratori. Se ne fa caccia. Cantore."

(Or, in English: "Blackbird. Very common. Resident; some individuals migratory. It is a hunter. A singer.")

So Napoleon's nephew saw resident blackbirds in Rome. Is
that really such a big deal? After rock pigeons and sparrows, the
sleek, sharp-billed birds (females: all-brown plumage and ditto
beak, males: black plumage, yellow beak and eye-ring; don't con-
fuse them with the American birds that go by the same name) are
probably the most abundant birds in cities—at least in Europe
and western Asia. In China and North America, close relatives,
the Chinese blackbird (*Turdus mandarinus*) and the American
robin (*Turdus migratorius*), take its place and behave pretty much
the same.

The reason that this short entry in Bonaparte's "Specchio" is
so pivotal is that it is the oldest record of blackbirds nesting
and wintering in a city that we know of. In the Bavarian cities of
Bamberg and Erlangen, blackbirds also frequented the town center in the 1820s, but they were not yet nesting there. And everywhere else in Europe, blackbirds at that time were still doing
what they had been doing for times immemorial: leading unob-
trusive lives deep in dark forests, so shy that they would be mor-
tified rather than be found in the company of humans, and, when
the breeding season was over, migrating down to the Mediterranean for the winter.

But, in the following two centuries all that changed. Slowly, at
first: by the end of the nineteenth century, urban blackbirds were
common sights only in the central parts of Europe. Throughout
the twentieth century, however, the city fad spread more rapidly,
reaching London in the 1920s and Iceland and parts of northern
Scandinavia only by the 1980s. Eventually, nearly every town
and city in Europe succumbed, except for a few small resistant areas in southern France, Russia, and the Baltic states. Over the entire

period, the urban tendency progressed at an average speed of five miles per year.

Still, this is not to say that blackbirds started to become urban in Rome and from there city-hopped their way across Europe. For a start, some of the blackbird's subspecies on islands in the Atlantic Ocean, far away from the species' stronghold in Europe, have independently made the same move into cities. Of their own accord, sometime in the middle of the twentieth century, *Turdus merula azorensis* and *Turdus merula cabrerae*, two smaller, darker and glossier versions of the blackbird that live only on the Azores (*azorensis*), and Madeira and the Canary Islands (*cabrerae*), also have begun hopping around in the islands' towns and cities. And the same happened even earlier in the north African subspecies *Turdus merula mauritanicus*: in the mid-nineteenth century, they were already city-dwellers in downtown Tunis. So, just like great tits across England becoming milk-bottle-top-savvy independently, all over Europe more and more cities obtained their own resident urban *Turdus merula* population.

Nobody is really sure why the city-dwelling streak spread among Europe's blackbirds so slowly but relentlessly. Why did it begin in the 1820s and not before or later, and what was it about Rome, Bamberg, and Erlangen that made those cities suitable a century earlier than, say, London or Brussels? And why do cities like Marseille and Moscow remain blackbird-free even today?

Obviously, deeper into the past, most cities may have been too small to constitute a viable habitat in their own right. Still, that cannot be the only answer: in the early nineteenth century, blackbird-free London already covered four and a half square miles, much larger than many smaller towns in Germany where blackbirds were already unflappably nesting in garden sheds and hopping around on the pavements. Parks and other

green spaces are important, but many cities had a generous helping of those long before blackbirds dared go there. Probably, the
growth of cities and urban green spaces, the balminess of urban
heat islands, a greater affluence of city-dwellers (leading to more,
and year-round, food surplus), as well as greater safety from hunters, predators, disease and parasites, all acted together to create
an urban niche snugly fit for blackbirds.

What *is* clear is that the process mostly happened in two steps.
First, blackbirds began wintering in a city. Then, sometimes only
many decades later, a few winter visitors stuck around for spring
and eventually took up breeding with one another, gave up migration altogether and turned into resident city-birds. Just like
what happened in the Californian dark-eyed juncos of the previous chapter.

That's about as much as one can glean from field guides and
birders' reports. To really get a feel for what makes the new urban
blackbird different from the ancestral forest blackbird, we have to
look at the work that an entire cottage industry of urban blackbird researchers has been doing over the past twenty years. Many
threads of urban evolution research that we have been following
throughout this book become tangled in this one urban bird.
The constellation of European cities has become urban evolution's
Galápagos, and *Turdus merula* its Darwin's finch. In almost every
European country, a team of urban blackbird biologists has
stepped up and jointly built a veritable urban evolution fest around
this one bird species—one of the world's oldest and best studied
urban animals. All this research seems to point in one direction:
that the urban blackbird is evolving into a separate species: a case
of true speciation.

We speak of speciation when a lot of different aspects of an animal or plant, either simultaneously or successively, evolve away

from the original type—to the extent that a taxonomist (the kind of biologist who circumscribes and classifies biodiversity) would consider it a different species. This usually means that its body shape, sexual strategy, and timing of major life events all begin to differ from the ancestor. In other words, when a complete over-haul of its genome takes place. But that alone is not enough. At least some of those changes must cause the original and the new gene pool to become separated and keep from blending into each other—something we call "reproductive isolation." (You can read much more about this in my book *Frogs, Flies, and Dandelions; the Making of Species.*)

Out there, in the wild, speciation often precipitates whenever a species colonizes a vacant, new niche. The shift in demands imposed by its new surroundings causes natural selection to en-force changes in its physique, its tolerances and behavior. When the ancestor of all Darwin's finches first landed on the Galápa-gos, there were plenty of unused niches: a whole smorgasbord of different plants and other types of food, whose nutritional bless-ings were just up for grabs. There were benefits to be had for any finch that was a little bit better able to target a particular diet—a process that, as we saw at the head of this chapter, is still ongo-ing. Because beak shape also determines the voice of a songbird, specializing on different food also brought about reproductive isolation: birds with different beaks sing different songs and no longer respond to those of birds with other beak shapes.

Cities are the world's new vacant niches, and the blackbird is one species that has embarked on the road toward speciating to maximize its profits from this horn of plenty, previously es-chewed by its reclusive ancestors. Let us briefly explore all the different manners in which urban blackbirds have become differ-ent, as discovered by those many blackbird teams across Europe.

(Think of them as the Blackbird Posse; there are too many teams, institutions, and persons to mention all of them individually. If you want to know the details, turn to the Notes.)

We begin with the most obvious but also the most confusing: their looks. At first, it seemed clear-cut. Researchers in the Netherlands and in France measured urban and forest blackbirds and found that the former had stubbier bills, were heavier, had longer intestines and shorter wings and legs. But when Karl Evans, a student of Kevin Gaston, looked at blackbirds from eleven cities all over Europe and North Africa, he found that this pattern did not hold everywhere. In some cities they had longer wings, in other cities shorter. For weight and leg length, the picture was similarly muddled. He did not check their intestines, but the only thing that *was* consistently different across all these cities was bill shape: everywhere, city blackbirds have shorter, stubbier bills than forest blackbirds, presumably thanks to the easy pickings at bird feeders and other places in the cities where food can be had without pecking, probing or pincering.

The Blackbird Posse has not yet investigated whether those different beaks have any effect on their voices, but for sure, urban blackbirds do have different songs. Male blackbirds possess a vast repertoire of melodious songs that they sing at dawn and dusk from high perches (branches and rocky ledges in the forest, TV aerials and rain gutters in the city). The score of each song, whether urban or otherwise, usually consists of a more or less elaborate motif, followed by a high-pitched twitter. And, just like most other songbirds in cities (see Chapter 16), the urban background noise forces the blackbird songsters to change their pitch and timing. As Hans Slabbekoorn's student Erwin Ripmeester found out after recording almost 3,000 songs, urban blackbird concerts are performed at a higher pitch than forest ones, and their twitters tend to be longer. And a German team discovered that, as foretold by Paul McCartney, urban blackbirds *are* singing in the dead of night. In the city center of Leipzig, they start a full three hours before sunrise, well before the trams and cars start creating a racket—whereas forest blackbirds open their beaks only at dawn.

Yet that is not the only thing that urban blackbirds do earlier. They also start breeding earlier in the year than their sylvan relatives. One of the causes for this is that their biological clocks are advanced by more than a month. In young urban males, the production of luteinizing hormone (which triggers the springtime flood of testosterone into their blood) peaks in mid-March, whereas for forest blackbirds, the call of spring does not reach fever pitch until mid-May. This was discovered by Jesko Partecke of the Max Planck Institute for Ornithology in Seewiesen, near Munich.

What Partecke did was raid ten nests of urban blackbirds in a cemetery in metropolitan Munich and ten rural blackbird nests in a quiet forest outside the city. From these nests, he took baby birds, thirty from each location, and brought them to his lab. There, he did one of those "common garden" experiments. As you may recall from earlier chapters, common garden experiments are a tried and tested way of making sure that differences between organisms are indeed genetic. In the case of blackbirds, the trick is to hand-raise the kidnapped chicks under identical circumstances and see what differences you still find. This way, Partecke could rest assured that the hormonal differences he discovered, at least the ones for young males, were not due to urban light or the urban heat island, or some other external trigger, but were purely the result of a genetically determined jolt to the urban body clock.

A second reason that the city blackbirds breed earlier is that they don't migrate. They spend the winter in the city, basking in their heat island and leisurely picking food off feeding tables, and can start breeding when they feel like it. The forest blackbirds on the other hand are, by and large, migratory: to escape the cold and the scarcity of food, they spend the winter in the south and only when they return to their home ground can they begin breeding. By that time, the blackbirds in the city are already smugly ensconced in their nests. And, as Partecke found out with those same hand-fed birds, the shift in migratory propensity is also genetic.

Since his birds were captive, he could not actually allow them to migrate away, so he did the next best thing: monitoring *Zugunruhe*. This German word literally means "migration restlessness" and it is ornithologists' and bird-keepers' jargon for the night-time jitteriness that befalls caged birds when their biological clock tells them to migrate, but the iron bars of their cage tell them otherwise. Partecke placed motion sensors in the individual cages where he held his blackbirds. This way, he could detect the signs of the onset of migration as indicated by the blackbirds' *Zugunruhe*. Sure enough, in autumn and spring, the forest birds were restless: at night, they moved about in their cage, hopping on and off their perches all the time. The urban birds, on the other hand, were fast asleep at night no matter whether it was migration season or not. Not only that, but the forest birds also built up massive fat reserves to sustain them during their anticipated flight, whereas the urban birds stayed lean. Curiously though, it was the males who showed this difference; the females differed much less between city and forest.

Partecke also discovered more fascinating urban/forest differences with his common garden blackbirds. For example, it turns out that city birds are much more laid-back in nature. This emerged when Partecke gave them the mildly stressful experience of taking them out of their cage and putting them in a cloth bag for an hour; at five time points during that hour, he opened the bag and took a blood sample. Measurements of the stress hormone corticosterone in the blood showed that the forest birds were much more alarmed by this procedure than the urban birds, whose stress hormone levels were raised only half as much as in the forest blackbirds. Let's remember that these birds had never seen a city or a forest; so urban blackbirds are really more even-tempered *by nature*. This may also explain why they allow humans to get three times closer to them before getting spooked than forest blackbirds.

One of the causes for this may lie in a gene called SERT, which stands for SERotonin Transporter. SERT is in charge of the removal of the mood-regulating hormone serotonin from the connecting points between nerve cells. That is why many antidepressants work by blocking SERT. As it turns out, forest blackbirds tend to have different versions of the SERT-gene than urban blackbirds.

So, that's a big pile of differences we find between urban and rural blackbirds: the way they look, their behaviors and personalities, their biological clocks . . . What does it all mean? In the wrap-up sentences of their scientific papers, the members of the Blackbird Posse exhibit their customary academic carefulness, but I'll just stick my neck out here and say it out loud: over the past centuries *Turdus merula* has spawned a new species, *Turdus urbanicus*, if you will. It's not quite there yet, just like those two Galápagos finches aren't quite there yet, but it's a matter of time for the process to be completed. It was only waiting for this moment to arise.

Not only does *Turdus urbanicus* boast a whole catalog of unique features, it's also puddling about in its own private gene pool. Blackbirds normally nest less than two miles from where they are born, so that already keeps their gene pools separate. And even if a forest blackbird would accidentally wind up in the city, it might be so ill-adapted that it would find it hard to get by. We know this thanks to failed attempts to introduce forest blackbirds into the Polish cities of Białystok and Olsztyn—whereas trials in Lublin and Kiev, this time using *urban* blackbirds, succeeded. Another reason why the urban genes stay in the city and the rural genes in the forest is that the urban birds start breeding so much earlier, before the forest birds have returned from their wintering quarters.

You can also simply have a peek at those gene pools themselves. That is what Karl Evans did. With a kind of genetic fingerprinting technique he tested the DNA of urban and forest blackbirds in twelve different sites all across Europe and North Africa. In all those places, the urban and forest blackbirds were genetically different, but it was also clear that everywhere, the urban birds had descended from the local forest birds. So there is not yet enough movement of blackbirds between cities (occasionally, blackbirds disperse much farther than those two miles) to homogenize them all into one urban gene pool. This is one of the main reasons why the Blackbird Posse is still a little reluctant to fully embrace the notion of a single, newly evolved *Turdus urbanicus*.

Still, the work of the Blackbird Posse makes an overwhelming case for the appearance of an urban-adapted evolutionary novelty. And surely it is not unique. We have seen many examples in this book of species that show a whole suite of freshly evolved genetic adaptations to urban conditions. And thanks to the urban heat island, if nothing else, many of those plants and animals will be flowering or mating earlier in the year than their relatives outside of the city. This alone may be enough to start the splitting of the gene pools, and incipient urban speciation.

This also means that evolutionary biologists no longer need to travel to remote places like the Galápagos to discover that holy grail of evolutionary biology: catching speciation in the act. The process is going on right in the very cities where they live and work!

But the reverse is, curiously enough, also true: in the Galápagos it has become possible to do urban evolution research. For today's Galápagos is no longer the deserted pristine place it was when Darwin first set foot there. Over 26,000 people live there, and the islands are visited by hundreds of thousands of tourists every year. The city of Puerto Ayora, on the island of Santa Cruz, where, if you recall, the Darwin's finch *Geospiza fortis* is in the process of speciating itself into two, has a population of 19,000 people, to which an annual deluge of 200,000 tourists is added. It has an airport, a highway (straight as a die), hotels, football pitches, tour agencies (*Natural Selection Tours*), cafés (*OMG! Galápagos*) and lots of restaurants.

It is those restaurants that, over the past few decades, the Darwin finches have begun frequenting. Thanks to their tameness (a characteristic shared by many island animals), it is no skin off their famed Darwinian noses to land on tables and feast on the morsels left by diners. And that—oh, the irony!—is beginning to undo the inchoate speciation process. Since the 1970s, the division between large and small-beaked Darwin's finches has begun to disappear in Puerto Ayora. Researchers such as Luis Fernando De León of the University of Massachusetts in Boston, who is studying those urban Darwin's finches, think this breakdown is due to their fast-food habits.

De León and his colleagues studied the feeding habits of urban and rural Darwin's finches and discovered that the ones in the city (where the two beak shapes have merged) eat mostly bread, potato chips, ice-cream cones, rice, and beans. They also drink water from the tap. The ones outside the city (where the finches still show two distinct beak sizes) feed, as they always did, on seeds of wild plants. Moreover, the urban finches also displayed the full checklist of urban bird personality changes, displaying curiosity when De Léon put out trays with unusual food items and being bolder at approaching humans making a "crinkle" sound with a bag of potato chips. The rural finches, on the other hand, showed no interest in humans or their food.

Thus are the ways of urban evolution. It giveth us a new blackbird species in Europe, and it taketh away a Darwin's finch on the other side of the globe. These and the many other examples in this book should make it clear that urban evolution is seriously reshaping our ecosystems. What does that mean for the future? How can we monitor or even steer this process? What roles could citizen science play? And could we perhaps even harness the power of urban evolution in nature-inclusive architecture and design?

IV.

DARWIN CITY

This natural beauty-hunger is made manifest in the little windowsill gardens of the poor, though perhaps only a geranium slip in a broken cup, as well as in the carefully tended rose and lily gardens of the rich.

JOHN MUIR, *The Yosemite* (1912)

19

EVOLUTION IN A TELECOUPLED WORLD

ONE OF MY FAVORITE CURES FOR writer's block (or, admittedly, sometimes, an act of procrastination) is my walk around the block. The centuries-old street plan of inner city Leiden, my hometown, is not actually divided so geometrically, so I take a more tortuous route. I descend from my writing den in the attic, and turn right upon exiting my front door and enter the Weddesteeg, the alley where Rembrandt was born. Taking slow and deliberate steps, trying to unfankle my mind, I cross the suspension bridge across the old, dead branch of the Rhine and turn left, and left again to cross the river once more, this time via the railway bridge. The embankment in between the pavement and the railroad is smothered in Japanese knotweed (*Fallopia japonica*), a relative of the rhubarb and, when it was first introduced into the Netherlands, highly prized for its spikes of white flowers, rich in nectar—large amounts were paid for the cuttings. Today, the knotweed's reputation has been tarnished by its unbridled spread and the ability of its roots to force apart brickwork and pavement. It is widely, but unsuccessfully, combated as an invasive exotic—in Leiden and nearly everywhere else in Europe, North America, New Zealand, and Australia.

My walk takes me back to the Rhine's left bank and then onto the Rapenburg, considered the Netherlands' prettiest canal, flanked by the lavish residences of its erstwhile notables. One of these homes of note is Rapenburg 19, a sixteenth-century colossus with plaster frills around the center doors and windows, dwarfing the more modest, delicately gabled houses on either side. There is a connection between this building and those Japanese knotweed plants that grow a stone's throw down the street. For this is the house where physician, botanist, ethnographer, and Japanologist Philipp Franz von Siebold settled after he had been kicked out of Japan in 1829.

Von Siebold is an interesting figure. Hugely famous in today's Japan (where he's fondly remembered as "Shiboruto-san" and the life story of his Japanese daughter Oine has been turned into a pop-culture manga), he was the only western explorer permitted into Japan during the more than two centuries of *sakoku*, Japan's policy to sever most ties with the outside world. Working as a physician from the Dutch trading post on the artificial islet of Dejima, off Nagasaki, he amassed a huge collection of local fauna and especially flora. He also collected ethnographical objects and, a crucial mistake, maps. Those maps were discovered by the authorities, he was accused of espionage, placed under house arrest and eventually forced to return to the Netherlands.

But he did not leave without his entire biological collection of dead and live specimens, which essentially was his pension. Back in Leiden, he managed to retire comfortably by selling off valuable parts of his collection, writing books, running a Japan museum from his home, and starting a mail-order oriental plant business founded on the live specimens he had brought with him. These included one living sprout of Japanese knotweed. It is cuttings from this one plant that kick-started the entire invasive world-domination of Japanese knotweed since. Those plants I passed on my walk along the railway bridge, a few hundred meters from Von Siebold's home, are its direct descendants, but so are the vilified knotweeds all the way across the world in New Zealand.

Besides knotweed, Von Siebold was responsible for spreading nearly a hundred other Japanese plant species to gardens and parks all over Europe and beyond. Eventually, some of those began to run wild. The *Wisteria*, Japanese rose (*Rosa rugosa*), Hortensia (*Hydrangaea macrophylla*), and garden privet (*Ligustrum ovalifolium*), now such familiar sights in urban green spaces everywhere, ultimately derive from Von Siebold's gardens. Even the Boston ivy (*Parthenocissus tricuspidata*), after which the "ivy league" is named, those prestigious ivy-clad American universities that house some of the urban biologists featured in this book, is originally a Japanese plant introduced to the rest of the world by Von Siebold.

Philipp Franz von Siebold set the scene. The nearly two centuries that have elapsed since he opened his Leiden-based Far-East flora shop have seen the birth of a million Von Siebolds. Global trade and commerce, garden centers, and pet shops cause the ever-increasing spread of animals and plants from their native homes to urbanized hubs around the globe. And then there is the unwitting transportation of seeds, fungi, microbes, and small animals in the clothes, luggage, shoes, and vehicles of the tourists, migrants, and other travelers who roam ever farther. Not to mention the teleportation of entire ecosystems by empty cargo ships who take in ballast water, soil, or rocks to improve their stability and then dump it near their port of call.

As I am writing this, I am on a two-month stint at Tōhoku University in the city of Sendai in Japan (home of the nutcracker crows). My writer's-block-laxative walks take me around the university campus and part of the city center. Here—naturally, since they are native plants—I also pass banks of Japanese knotweed, festoons of *Wisteria* blossoms, bushes of garden privet, and buildings clad in Japanese ivy, just like I do on my jaunts in Leiden. But I also come across familiar European plants that have made the voyage in the opposite direction of Von Siebold's transportations. In the poorly tended lawns near the Kawauchi subway station I spot shepherd's purse (*Bursa capsella-pastoris*) with their funny-looking seed pods, and also white clover (*Trifolium repens*). In the roadside verges of Jozenji Street grow garden sorrel (*Rumex acetosa*) and the common broom (*Cytisus scoparius*). And not only plants, either. Above the city fly rock doves, while buff-tailed bumblebees (*Bombus terrestris*) buzz their way through the beds of white clover flowers. On wet nights, the European slug *Lehmannia valentiana* slithers across the walls of the Shintō shrines, as comfortably as if it were at home. In fact, the urban habitats of Leiden and Sendai are much more similar, share many more species, than was the case in Von Siebold's days.

I also see some true cosmopolitan invaders, of the kind that ecologists call "supertramp species": in the fields on campus grows tall fescue grass (*Festuca arundinacea*), originally from Europe, but now found on any lawn between the two poles (including the south lawn of the White House). And the walls of the moist bathroom in my rented apartment show the orange coloration of *Aureobasidium pullulans*, a fungus that forms one giant gene pool all over the world, continuously mixed by people carrying their toiletries from bathroom to bathroom.

Such global homogenization of urban ecosystems is much more pervasive than these few anecdotal examples suggest. Urban ecologists worldwide are doing inventories that show clear evidence that unseen hands are causing cities worldwide to become communicating vessels for all manner of organisms. For example, an international scientific consortium that calls itself GLUSEEN (Global Urban Soil Ecology and Education Network) has recently carried out DNA surveys of soil microbes in cities and natural areas in Africa, North America, and Europe, and discovered that the species compositions of city soils across those continents are converging. At least for the 12,000 species of fungi and 3,700 species of a type of microbe called Archaea they found, the communities underfoot in cities were much more similar than in forests. And a team of researchers from the University of New Mexico used the information gathered by the nationwide American tradition of the Christmas Bird Count, where citizen scientists in thousands of locations count all birds in a 15-mile-diameter-circle in 24 hours. They discovered that cities as far apart as 2,500 miles still share about half their birds, while the avifaunas of *natural* areas that distant are almost complete different. Two German researchers, Rüdiger Wittig and Ute Becker, finally, did an analysis of the plants found growing around what they call *Baumscheiben* (tree disks), the small islands of soil around the base of street trees. Just like in the bird studies, a much greater proportion of the *Baumscheiben*-flora was shared by cities all over the continent, than by random *Baumscheiben*-sized plots in natural vegetation. Even in the American city of Baltimore, 80 percent of the herbs growing at the roots of street trees were identical to those found in the European cities.

What all this means is that the ecosystems of cities around the world are growing more and more alike; their communities of plants and animals, fungi, single-celled organisms, and viruses are slowly inching toward a single globalized, multi-purpose urban biodiversity. And even if the exact species across cities may not be identical, you will find similar species playing similar roles. On my walks in Sendai, for example, I see the spider *Larinioides cornutus* weaving its webs near lights on bridges across the Hirose river. This is a different species from the European bridge spider (*Larinioides sclopetarius*) that Astrid Heiling observed doing the exact same thing on Viennese bridges across the Danube earlier in this book.

The upshot is that each urban species, wherever it finds itself in the world, will encounter quite a similar set of urban cohabitants. In a previous chapter, we called these "urban encounters of the second kind": the living cogwheels of a city's ecological clockwork that can themselves also evolve. As those wheels in different cities converge, this will lead to similar evolutionary breakthroughs for comparable struggles for urban life all over the planet.

But there is also homogenization in what I called earlier, "urban encounters of the first kind": adapting to the physical and chemical signatures of the city environment. For those components of the urban environment, there are global, invisible connections as well. Over recent years, the visionary urban scientist Marina Alberti of the University of Washington has been revealing this process, which she calls "telecoupling." As I interview her over Skype she first explains where she is coming from: "Yes,

I am an urban planner, but my background is complex and I have studied biology as well." She is rooted, she says, in the idea that humans are fully part of nature. "My work has tried to challenge both ecology and urban planning since both still see humans as separate from ecological systems."

She continues: "Cities are networked far beyond their own physical edges." What she means is that cities exchange not only species, but also the human inventions that make those cities tick and that urban organisms must adapt to. Take artificial lighting at night (ALAN), for example. Over time, one light innovation after another has swept across the world, leapfrogging from city to city. First there was gas light, then incandescent lamps, followed by high-pressure sodium, mercury vapor, and now, LED lights. Each of those types has a different light spectrum, and ecologists such as Kevin Gaston in the UK and Kamiel Spoelstra in the Netherlands are finding out that nocturnal animals respond differently to different spectra. So, if animals are indeed evolving to deal with ALAN, each lighting innovation will cause a new swing in their urban evolution. And since those technological innovations will spread rapidly from city to city, so will the evolutionary swings, either because the better-adapted animals themselves disperse between cities (like the pollution-adapted peppered moth did) or because the same solutions for the same problems evolve independently in different cities (like what seems to be going on in the city blackbirds, and in PCB-tolerant mummichog fish).

Alberti envisages that such technological telecoupling among cities will apply to any innovation in transportation, road and train construction, architecture, green space planning, et cetera. All the more so, because cities worldwide are beginning to exchange information and undertake concerted action more enthusiastically even than the countries they are in. In his book, *Connectography: Mapping the Future of the Global Civilization*, global strategist

Parag Khanna gives the example of the C40, a global network of megacities that work together to invent and implement solutions for climate change. "Because cities define themselves [...] by their connectedness rather than their sovereignty," Khanna writes, "one can imagine a global society emerging much more readily from intercity relations than international relations."

In other words, urban nature of the future could be a globally homogenized, dispersed ecosystem, inhabited by a dynamic but shared set of organisms that is constantly evolving, exchanging species and genes and innovations to deal with the new technologies with which humans equip their cities. Not that the urban ecosystems will ever be completely cut off from natural ecosystems: the wild will keep functioning as a source of pre-adapted species and genes that urban ecosystems may put to good use. Still, as the urban environment expands its reach, it will become more and more an ecosystem in its own right, writing its own evolutionary rules and running at its own evolutionary pace.

These urban evolutionary rules and pace, Alberti points out, are beginning to diverge more and more from the ones we're used to in the natural world. There, in the unspoiled forests, deserts, swamps, and dunes far from human interference, evolutionary change is driven by age-old natural forces. As wild ecosystems get larger and more complex, and niches get filled, fewer and fewer opportunities are left to exploit and evolution may slow down. But in cities, says Alberti, it is the other way around. The pace of evolution is set by ecological opportunities that emerge from human social interactions. The bigger and more complex cities get, the more intensive human social interactions will be, and the more rapidly the environment will change as a consequence—and telecoupling then ensures that these changes reverberate through the network of the world's cities. In those pressure cookers of environmental change, species will need to speed up their evolution or become extinct.

Some species, whose evolution cannot keep up with the frenetic rate at which the urban environment develops, will indeed drop out, but others will continue adapting and perhaps splitting into multiple species as humans pose them new obstacles or offer them new ways of living. In 2017, in the journal *Proceedings of the National Academy of Sciences*, a team of authors led by Alberti published a global analysis of more than 1,600 cases of "phenotypic" change (an alteration in the appearance, development, or behavior of a species, which may or may not be genetic). When they factored in a whole set of environmental "drivers" (some urban, some natural), they discovered what Alberti calls "a clear signal of urbanization." The data showed that phenotypic changes in cities happen faster than those farther away from cities, and that the strongest drivers are encounters of the second kind (interactions with humans themselves or with other organisms brought into the city by humans).

That leads us to a question that I alluded to in an earlier chapter: what about ourselves? Could we be evolving as well? After all, the urban environment is just as foreign to our bodies and minds as it is to house crows, knotweeds, and mummichogs. We have never lived under conditions even remotely like it for the hundreds of thousands of years that have shaped our evolution until now. Moreover, if there ever was a time that could produce the raw material for a new spurt in our evolution, it is today. Think about it: with nearly eight billion people on earth, and a quintillion sex cells produced in our bodies on a daily basis, there is a greater chance on new, crucial mutations appearing in our genomes now than there was when we still were an endangered species eking out a living in a few forgotten corners of the world.

Are we humans perhaps evolving to suit our new urban environment just like all the animals and plants that we cohabit with? It's an intriguing question, and one that we now have the tools to answer.

Just like scientists scanned the genomes of peppered moths to pinpoint the mutation in the *cortex* gene that swept across the industrially revolutionized world in the nineteenth century, we can scan the genomes of an ever increasing number of people to detect how the human population has evolved in recent years. DNA technology is developing so quickly that it is only a matter of years before everybody will have their full genome as private information on a hard-disk and/or allow researchers to mine it for scientific information. But even with the modest amount of one million or so human genomes that researchers have access to today, they are picking up signals of ongoing human evolution.

The UK10K and Biobank projects, for example, two UK-based initiatives to sequence the genomes of tens of thousands of Britons, have revealed that recent centuries have seen evolutionary changes in genes that link with height, eye and skin color, lactose tolerance, nicotine craving, head size in infants, hip circumference in women, and age at sexual maturity in women. Probably none of these evolutionary shifts has to do specifically with living in cities. But other researchers *have* found evidence for city-specific human evolution. For example, in parts of the world where urbanization began earlier, more people carry immune system genes that help fight tuberculosis. Organisms that cause or spread such diseases do so more easily where people are densely packed—even in today's modern cities—and that will affect the evolution of our immune systems.

Another titillating idea is that human sexuality will change. For millions of years, any human would meet only a few handfuls of possible partners in his or her lifetime. The average urbanite of today, however, runs into that many potential candidates during a walk around the block. That means more competition and a much more intensive sexual selection. Combined with perhaps a great-tit-like shift in perceptions of what makes a perfect (urban) mate, who knows how our sexual signals and preferences are going to evolve in the future?

Having said that, for the foreseeable future, it is more likely that urban humans will be influencing the evolutionary trajectories of urban flora and fauna rather than the other way around. Alberti: "I think that the human species is changing the genetic makeup of the planet. We have both the responsibility and the opportunity to co-evolve with other organisms. Whether humans will take this challenge, I do not know." Alberti formulates a challenge with great implications for how we are going to design and manage our urban environments. Can we harness the power of urban evolution and use it to make more liveable cities for the future?

20
DESIGN IT WITH DARWIN

MY DAUGHTER IS MUCH BETTER AT this sort of thing.

I show up unannounced at the reception desk of Roppongi Hills, a grandiose "integrated property development complex" in the heart of Tokyo, asking if I can have a look at their famous green roofs. I get a glossy *Roppongi Hills Town Guide* and a cascade of apologies, but, no, very sorry, they are only accessible during "events" and special occasions. Despondently, I return to my daughter and girlfriend who are waiting outside. "But did you tell them you're working on a book?" my daughter Fenna asks, and pushes me back inside. Her smiles and persistence, and my conjuring a collection of business cards and university IDs eventually result in the desk clerk phoning the company's greens management office, but they, too, politely turn us down. Book or no book, you must file a special request.

But Fenna is in a belligerent mood. "Let's see how close we can get to it," she suggests, and ushers me into the lobby of the neighboring Hyatt Hotel, where we sneak past the reception desk and into the elevator. For a few minutes we shuttle between floors until we find the one that approaches closest to the level where we expect the Roppongi Hills green roof. At the end of the carpeted corridor in which we find ourselves is a large window, which, as we approach it, opens up to a great vista. First, the hazy, jagged skyline of Tokyo comes into view. Then, the "town in the city" that is Roppongi Hills: a cluster of offices, shops, apartments, gardens, and a museum, interconnected by sculpture-lined paths. And, as we press our noses against the window and feel the urban heat radiating through the glass into the air-conditioned hotel air, there, right in front of us, in the blazing sun, lies the rooftop urban nature reserve we had been looking for.

It's as if a chunk of the Japanese countryside (that traditional mosaic of rice fields, forests, grassland, and ponds known as *satoyama*) has been dug up and plonked on top of the Keyakizaka building, where it sits like a green crown. A rice paddy and several meadows are fringed by cherry woodland and privet hedges. Walkways snake through them. There are lotus flowers in a pond and vegetable plots with bitter gourd, eggplant, and tomato. A resident in a sun hat is tending to her rice, while two jungle crows with bills like sushi knives are nibbling at the young cherries of the cherry trees. Then they take wing and flap toward the gigantic Mori Tower, looming high overhead. We watch them get smaller and smaller until eventually they are two dots that dissolve among the ledges of the top floors, where the Mori Building Company, who designed Roppongi Hills, has its headquarters.

Since the early 1970s, Mori has been one of the leading building firms to incorporate vegetation into their architecture. The rooftop garden at Roppongi is modest, at only 1,500 square yards, but some entries in their portfolio have much larger surfaces covered with green. And they are not the only ones. In the city of Fukuoka, the Argentinian designer Emilio Ambasz created the project developer's equivalent of having your cake and eating it: he took a 120,000-square-yard city-center park and effectively lifted it up into the air, sticking his futuristic Fukuoka Prefectural Hall underneath. The south-facing slope of the wedge-shaped building is covered in fourteen terraces of wild-looking vegetation that merges with a ground-floor park at its foot.

Elsewhere in Asia, there are other record-breaking green architecture projects. Singapore boasts both the largest vertical garden (the 2,750-square-meter west-facing wall of the CDL Treehouse Condominium) and the stunning, thirty-story creeper-covered Oasia Hotel Downtown. Not that Asia is in any way unique in the green building world. In Milan, green architect Stefano Boeri built the *Bosco Verticale* (vertical forest), two residential towers planted with 730 trees, 5,000 shrubs, and 11,000 herbs. In fact, nature-inclusivity is in full swing in urban design and architecture worldwide, and more and more clever, ecologically inspired ideas are being developed to integrate nature into the city, on megalomanic as well as microscopic scales, and by building tycoons as well as tiny start-ups. In Manhattan, for example, the crowdfunded Lowline Lab is experimenting with creating underground green spaces under poor light conditions. The aim is to convert the abandoned, 200-yard-long Williamsburg Trolley Terminal underneath Delancey Street into dank

cavernous spaces where mosses and ferns may thrive in a subterranean park. And in Berlin, a local community is in the process
of converting a behemoth of a Nazi-era concrete bunker into a
"green mountain," christened *Hilldegarden.*

These new trends in design do not just give a boost to the ar-
chitectural and designers' guilds, there are many other benefits
for the urban environment. For example, roofs are blatantly vacant
in areas where competition for space is cutthroat. With cities growing and a rapidly dwindling availability of flat, low-lying land for
nature and agriculture, why not bring these land-uses into the city,
and onto those unused roofs? An added advantage of vegetation-
clad constructions is that all that humid soil and foliage help a
building stay cool. Air-conditioning costs go down and the urban
heat island is mitigated as well. Moreover, the plants muffle noise
and trap smog. In earthquake regions like Japan, a heavy rooftop
park may even act as a counterweight and help to quake-proof the
building. No wonder that in 2001, Tokyo passed an ordinance that
says that new buildings need to devote 20 to 25 percent of their
roof surface to greeneries. By 2015, this regulation had already
yielded a total of 2.2 million square yards of green roofs. Similar
regulations and incentives exist in cities all over the world.

Now, I will not give you an in-depth overview of the big and
booming bandwagon of green urban design and architecture.
There is already an entire library about this, and I could refer you
to the excellent *Making Urban Nature* (2017), and *Designing for
Biodiversity* (2013). Or the classic *Planting Green Roofs and Living Walls* of 2004. Or, you could attend one of the many international conferences where architects and planners congregate
to discuss the latest developments in ecological urbanism. But
what you won't read in those books or hear in those presentations
is how all these new trends in green roofing, nature-inclusive
building, and green urbanism affect the ongoing evolution of

urban ecosystems. By and large, urban ecologists, ecological architects, and green planners assume that the animals and plants that they release in the urban environment are static—that the roles they play in the network of city life will remain unchanged. As we've seen in this book, this may be a miscalculation. There is always urban evolution, and the question is: what's the cross-talk between urban evolution and urban architecture?

First of all, there's the telecoupling that we talked about in the previous chapter. Urban engineers, designers, and architects engage in intensive global exchange of ideas. The Argentinian architect Emilio Ambasz creates buildings in Japan, and the Italian Stefano Boeri does the same in China, while Malaysian Ken Yeang works on projects in London, Hong Kong, and Bangalore. All this mobility of people and concepts ensures quick global spread of insights and innovations, and this adds to that telecoupled global urban environment of which we spoke.

But there is more. With a little evolutionary reasoning, it may be possible to come up with a few guidelines on how urban design may harness the power of evolution to assist in the evolutionary maturation of those urban ecosystems. So, here come my off-the-cuff Guidelines for Building with Darwin: four rules for evolutionarily informed urban nature planning.

Number 1: *Let it grow.* We humans are incorrigible gardeners. We want to plant, to weed, and to arrange. And green urban design is no different. All the greening projects that I mentioned at the start of this chapter, whether vertical, horizontal, slanted or underground, are meticulously planned—not just in looks and function, but also in terms of what shall grow there. The Fukuoka Prefectural Hall roof garden was seeded with seventy-six species of herbs, bushes, and trees. The Lowline Lab in New York used more than a hundred, and Milan's *Bosco Verticale* boasts fifty different, carefully selected types of plant. In each of these projects, a team of horticulturalists and arborists would have concocted the ideal mixture of species to suit the specific environment. They would have combined properties such as heat, shade, and drought tolerance with aesthetic characteristics of the forms and colors of leaves, stems, branches, and flowers.

While the hand-picking of such elite troops is understandable, it totally ignores the motley crews of urban plants that these new green spaces are parachuted into. Everywhere in the city, in gutters, roadsides and on non-designed rooftops, communities of plants are co-evolving with each other, with the micro-organisms in the soil and the air, with the insects and other invertebrates that eat and pollinate them, and with the urban environment (the heat island, the patchiness of the soil, the heavy metal pollution, and so on). These evolutionary processes are not helped by dropping a foreign body of pre-assembled plant species among them. Much better would it be to let the green spaces assemble naturally from species growing abundantly elsewhere in the city. This would entail not planting anything, perhaps not even adding soil, but simply leaving the beds empty and letting the urban ecosystem colonize it under its own steam. On a (very) small scale, there are already initiatives that adopt such a "let it grow" philosophy. The Dutch company *Gewildgroei* (an untranslatable pun meaning something like, Desirable Growth) designs and distributes pavement tiles that contain gaps and holes for soil to collect and plants to sprout spontaneously. On the large scale of an entire building, however, such a laissez-faire approach would mean that that glitzy new "green" project would look terribly bare for the years leading up to the eventual autonomous greening.

Number 2: *Not Necessarily Native*. If we cannot bear the sight of a naked "green" building, and we really must inoculate it with trees, bushes, and herbs, then why not choose from the local urban portfolio? Any green architecture project will find itself in a city that is already full of urban plants, evolving and adapting to that particular urban environment—ideal starting material! Explore vacant lots, rooftops, and railway embankments and select from the plant species that are already thriving there. "But wait," the ecological architect would object, "many of those species would be exotic—wouldn't that go against the urban greening creed to use only native species?" Well, yes, it would; and I maintain that such a creed would be misguided. Of course, it sounds cozy to plant native flora in the city, but we have to face the fact that many of the species that have been evolving and adapting to the urban environment most successfully are non-native— we have come across quite a few examples in this book. It is those ecological supertramps, those citizens of the world that will make up the bulk of that globalized urban ecosystem, and urban planners could do worse than yield to that inconvenient truth of urban evolution.

Number 3: *Pristine Pockets.* To preserve pockets of original, non-urban habitat inside the city perimeter may seem to go against what I suggested just now in *Not Necessarily Native*. Still, to keep the engine of urban evolution running, it is important to have a large reservoir of species and genes handy for ecological innovation. The evolving urban ecosystem will regularly be faced with new challenges, and not all the species of the urban food web will be up to the task of adapting to the next urban novelty. That's why pockets of natural vegetation, which still retain the original, local flora and fauna, can act as a safety valve. Places like the Hassamu nature reserve in Japan's Sapporo, the Campus do Pici forest in the Brazilian city of Fortaleza, and Bukit Timah in Singapore are such remnants of original old growth forest woven into the fabric of a metropolis.

Number 4: *Splendid Isolation.* One of the central tenets in urban green design these days is "corridors." Creating linear strips of vegetation ("greenways") between parks and other fragments of vegetation in the city to make an interconnected network of urban green spaces is all the rage. It seems like a good idea. After all, it is the urban equivalent of what has been standard practice in nature conservation outside of the city for many decades. When a species disappears from one fragment, it could recolonize from another. This way, the food webs in all those networked reserves stay intact.

Whether corridors are always a good thing in the evolving *urban* ecosystem is another matter. Think of those white-footed mice from New York, each clan adapted to the specific demands of the park that it found itself isolated in. To them, it may actually be a good thing to be trapped in their own park and not be blended all the time with poorly adapted mice from other parks. The same may be true for many other smaller animals and plants that are trapped in the smallest of urban pocket parks. Like those mice, they probably evolve to adapt to the idiosyncrasies of their particular corner of the city. Connecting those corners via corridors will link those populations and break down those delicate adaptations. So, for the evolution of much of urban life, it might be wise to think twice before planning a corridor.

As you will have noticed, some of these guiding principles go against the grain of ecological urban planning dogma, and it may take a while before city councils internalize a sense of urban evolution. If one has spent decades combating exotic species in one's city, it will be hard to own up to the fact that one's efforts are only delaying the inevitable process of evolutionary integration of those species. Similarly, the notion that some urban evolution may be more successful in small, isolated pockets of vegetation than if you connect those pockets with green corridors could be hard to swallow.

So, perhaps we should not count only on the authorities to

take a lead in evolutionarily enlightened urban ecosystem management. Communities of concerned urbanites form perhaps a more powerful vehicle to make a change, and in many countries, city dwellers are already grouping around a shared interest in their local nature. Tokyo is a case in point. As the city grew from one million people in the late seventeenth century to today's urban conglomeration of 38 million (the Japanese capital is still in the world's top three, after Guangzhuo and Chongqing), its outskirts swallowed up the surrounding *satoyama* countryside. *Satoyama*, as is explained to me by urban ecologists Tetsuro Hosaka and Shinya Numata of Tokyo Metropolitan University, is a catch-all phrase that encompasses the way rural people in Japan have traditionally been managing their natural environment, integrating it in the patchwork of villages, agricultural fields, irrigation channels, and coppice woods.

But, says Hosaka, while those landscapes may be gone, people's desire to work together toward their common—green—good is still there. Over recent decades, this has led to a reincarnation of the *satoyama* concept: city dwellers get together and jointly restore and preserve *satoyama* landscapes on the city's fringe. There are even some initiatives to apply the satoyama concept to the city center, with groups of neighbors managing the parks, canals, ponds, and roadsides in their quarter, just like their ancestors did in pre-urban times. Moreover, urban farming also becomes a part of this city *satoyama*, particularly among the sizeable sixty-plus demographic. Numata: "Some old guys are very enthusiastic because they like gardening and vegetable plots. There are so many retired but still powerful and resourceful people. In today's Japan, sixty-year-olds are still very young and full of energy." Indeed, on my train journey toward Tokyo Metropolitan University, I see plenty of evidence of this. Everywhere, little orchards and vegetable gardens, even miniature rice paddies are tucked away among the gray apartment blocks. Even in the heart of the city, the affluent residents of glitzy Roppongi Hills come together to plant, tend, and harvest their rice on that roof of the Keyakizaka building.

And Japan's urban *satoyama* revival does not stand on its own. All over the world, from Amsterdam to Acapulco, and Zamboanga to Zhengzhou, local urban communities are taking up neighborhood nature conservation and urban farming. The importance of the farming component should not be underestimated, because it means that humans become an integral part of the urban food web. Through their farmed fruit and vegetables, urban humans and their digestive systems are taken up in the energy flows of the local ecosystem. And, since it concerns their stomachs and their well-being, they will automatically become more concerned about that ecosystem.

Perhaps, such groups of ecologically minded city dwellers (which partly overlap with those urban naturalists we came across at the beginning of this book) could prove to be a fertile ground for the notion of urban evolution. In fact, many of the scientists I interviewed for this book told me that city folk are thrilled when they hear that urban flora and fauna are actually evolving to suit the city environment. Think of the potential of involving citizens into urban evolution research!

And that brings me to the final point I would like to make: Let's build ourselves an Urban EvoScope!

Urban evolution is everywhere. All the animals and plants in our cities are rapidly changing and adapting to those cities. But, with the exception of the handful of urban evolutionary biologists that have marched across these pages, nobody is watching this. There simply aren't enough scientists to constantly monitor the changing chuckwallas of Phoenix, the evolving eagles of Vancouver, or the adapting adzuki of Shanghai. But, with a global urban population of 4 billion people, there might just be enough citizen scientists to do the job. What if some of those urban citizen scientists, instead of simply recording the presence or absence of species, would also record their evolution?

To give you an example of what I have in mind: in the Netherlands, we recently launched a smartphone app called Snail-Snap. With it, people can upload photos of the grove snail (*Cepaea nemoralis*), common in all Dutch cities, to a central database. The shells of these snails come in many different colors, and our team uses the thousands of photos to figure out whether the shells in the city center are evolving toward lighter coloration. The idea being that snails in a bright shell sitting in the mid-summer urban heat island may overheat just a little less quickly (and therefore survive a bit better) than snails with a dark shell.

Another example comes from acoustic ecology. You'll recall the changing sounds of birds, insects, and frogs that try to communicate in the urban noise. There are several projects around the world in which volunteers place very small USB-microphones in their gardens or on their homes' outside walls that automatically, and continuously, record the local "soundscape." In some cases, like the network of detectors installed in London, this even includes the ultrasonic bandwidth, to monitor bats' sounds. This way, acoustic ecologists can monitor how animals' calls, songs, and stridulations change as a result of the racket we humans create.

Then there are some fun ongoing projects that could also be used to keep track of changes. For example, the so-called Funky Nest Contest, run by the Cornell University Laboratory of Ornithology, gets people to send in pictures of the funniest, most adorable, inconvenient, bizarre . . . well, the *funkiest* birds' nests they find in their urban environment. This sort of project could reveal changing (maybe, evolving?) nesting behavior. Do urban blackbirds indeed nest more in and on artificial places than you'd expect by chance? What's up with those white plastic strips that urban black kites in Spain have started decorating their nests with? Do other birds perhaps also line their nests with cigarette butts like house finches in Mexico do?

These are just a few existing ideas, but think of the possibilities when technologies improve. In the not too distant future, when DNA analysis devices become small and cheap enough, citizen scientists could be monitoring the actual changes in the genes of the urban animals and plants. With improved image recognition software, the photos uploaded to citizen science websites could be used for tracking changes in insect coloration, seed shape, leg length, and all those other ways in which urban flora and fauna are evolving. Together, these monitoring schemes could become a global, permanent EvoScope that keeps tabs on the fluid Darwinian motions that every city ecosystem goes through.

OUTSKIRT

OKAY, THIS HURTS. FOR YEARS, I HAD avoided coming here. But today, after visiting my mother who still lives in the same 1950s house where I grew up, I decided it was time to take a walk through the new suburbs that have sprung up in the place of the fields and swamps on the edge of Rotterdam, where, as a teenage boy, I laid the foundation of a lifetime of nature study.

It is literally gut-wrenching. I walk among rows and rows of semi-detached houses of the kind that Dutch town planners have churned out by the millions. They are pleasant enough: child-friendly neighborhoods with cute gardens, carports, winding streets with speed bumps, cozy-sounding street names. But to me they are a mausoleum for my old stomping grounds. It is only with the aid of Google Earth that I can work out where, in my outdated mental map, I am.

The neighborhood of De Akkers (a prefab garden shed, an affordable car, and an Ikea umbrella in every front yard) is where I was enveloped in a flock of hundreds of common snipe (*Gallinago gallinago*), circling the morass while making their sneezing noises in unison—a memory that has stayed with me ever since. Where the De Velden apartment building now stands, I lay in the tall grass with my cheap telephoto to take pictures of nesting black-tailed godwits (*Limosa limosa*). Now, ring-necked parakeets dangle from strings of peanuts at a bird feeder. And the soil underneath De Gaarden bus station is roughly where I caught the subterranean beetle *Choleva agilis* in the burrow of a common vole—the pinned specimen now sits in the collection of Naturalis, the natural history museum where I work.

The semi-natural landscape of bogs and pasture where I roamed no longer exists. It has been subsumed into the sprawl of Rotterdam and converted into the kind of urban environment that I have waxed lyrical about in this book. And yet, the confrontation with the starkness of this process of conversion makes me sad. Is that inconsistent of me? No—of course we must regret what we lose, but that does not mean that what we gain is worthless.

The landscape of my youth instilled sadness in my grandfather, who grew up in the same area at the start of the twentieth century, in a time before pesticides and fertilizers, when insects and wildflowers were much more diverse and plentiful than in "my" 1970s. And to the children growing up today in this suburbia, the remaining ditches, grassy strips, and hedgerows in between the buildings will form the backdrop for *their* childhood memories, which will be just as precious to them as mine are to me. In other words, as our human footprint gets bigger, the natural world that surrounds us shrinks, changes and becomes poorer. But, biologically impoverished as they may be, these urban ecosystems are still exactly that: ecosystems, with real organisms, suspended in real food webs where real ecology and real evolution go on.

Natural selection here is so strong that urban life forms evolve rapidly. But we must also remember that all the examples of urban evolution in this book form a biased sample of those life forms that were pre-adapted, variable, or simply lucky enough to evolve and survive. For each successful urban species there are dozens of other species that could not adapt to city life and disappeared. Besides being evolutionary powerhouses, cities are also places where great loss of biodiversity takes place. No matter how interesting they are biologically, we cannot rely on them for the preservation of the bulk of the world's species. For that, we must preserve, appreciate, and explore what remains of pristine, unspoiled wilderness.

With our organization Taxon Expeditions, for example, Iva Njunjić and I take eco-tourists on expeditions into the unspoiled rainforest of Borneo, where we discover and name entirely new species of wildlife. But most people will never go into the jungle. All they will see of nature for most of their lives will be that neighborhood park, or those few plants and insects in their back garden. That's why it is so important not to dismiss these bits of urban ecosystem as dreary and uninteresting. That's why an awareness of the exciting evolutionary processes going on in cities is so crucial for the quality of urban life.

I hope that, by reading this book, your eyes have been opened to the wonders of HIREC, Human-Induced Rapid Evolutionary Change. One of my aims is that the urban organisms you see on your daily wanderings of the city streets will now become more special, more interesting, worthy of more than a casual glance. So when you see a flock of pigeons, you'll look out for the ones with dark feathers and think to yourself, "Hey, those are the ones that can deal better with the zinc flaking off that lantern post over yonder." That, when you see insects circling the fluorescent tube at a vending machine, you'll imagine that it's actually the ones genetically predisposed *not* to be drawn to that light that are the chosen ones for future urban insect life. That, when a blackbird crosses your path, you will realize that this species is your city's answer to the Galápagos finch. And I hope that, over the years to come, there will be citizen science projects for you to observe evolution in action where you live. Or, better still, that you'll start one of your own.

What will the future bring? At least in the short run, cities and urban populations will grow even further and we'll claim an even bigger role in the world's food web. Sometime in the twenty-first century, probably half of the energy that the ecosystems on earth produce will, directly or indirectly, pass through us. In ecology, a species with such a central role is termed a keystone species. Humans are a keystone species of unprecedented magnitude: we are a hyperkeystone, ecosystem-engineering supertramp species.

As you may remember from the ants and their myrmecophiles at the beginning of this book, powerful ecosystem engineers are like a magnet to other species. With such concentrations of food and resources, other species evolve to cohabit with them—for shelter and protection, to steal and pilfer from their hosts, or to trick them into giving them their scraps. Many go unnoticed, some are tolerated or even valued, and some prosecuted, but all evolve, adapt to a life shared with their willing or unwilling benefactors. Humans have been in this position for a much shorter duration than ants, and our anthropophiles are still only starting to evolve. But evolve they do, and further adapt they will.

Especially if we give them a helping hand. By observing, monitoring, and understanding urban evolution, we can design our urban environments in such a way that we may harness and steer this process. We can engineer our own ecosystem engineering. And we should do so in a constructive way, by applying Darwinian rules to urban greening—not in a destructive way by weeding out those species that actually hold the best cards for evolving into anthropophiles. Like the house crow.

Once again, I contact Sabine Rietkerk of the Save the House Crow committee. It has been nearly a year since I last was in touch with her, and things are not well in Hoek van Holland, she tells me in rapid, staccato Facebook messages. The hunters have returned and, in early spring, have managed to kill the last remaining birds. And that includes the one I saw sauntering across the street in the shopping center. "He lasted the longest," Sabine writes. "He was always the one who was most alert, always calling, always warning the others of danger." It is almost as if he threw in the towel when there was nobody left to warn.

So, the potential urban evolution of a North European version of the tropical *Corvus splendens* has been nipped in the bud. "Well . . . There's still that rumor," she writes. Which rumor? "That somewhere in the neighborhood, a few crows have gone into hiding in somebody's home." Is that true? I ask her. "That's the neighborhood's best kept secret," she writes back.

And then, after a few seconds, Facebook messenger pings one more time: ";-)."

NOTES

CITY PORTAL

The descriptions of nature in inner city London was based on a visit to London June 21–24, 2016. The story of the London Underground mosquito was based on Shute (1951), Byrne & Nichols (1999), Fonseca *et al.* (2004) and Silver (2016), and on the presentation by Katharine Byrne at the European Society for Evolutionary Biology congress in 1995 in Edinburgh. The data on the history of urbanization are from Merritt & Newson (1978), Seto *et al.* (2012), Newitz (2013) and Reumer (2014). The recent study showing the increasing distance to the nearest forest is Yang & Mountrakis (2017). The amount of primary productivity that humans appropriate is from Imhoff *et al.* (2004) and Haberl *et al.* (2007), and the amount of freshwater consumed is from Postel *et al.* (1996). The opinion paper I mention is Huisman & Schilthuizen (2010). The area where I spent much of my childhood naturalist days in the late 1970s and early 1980s were the fields and swamps north and northwest of the village of Kethel, part of the municipality of Schiedam. Much of that area was converted to residential areas in the 1990s and the early twenty-first century.

CHAPTER 1: THE ULTIMATE ECOSYSTEM ENGINEER

The beetles I collected in Voorne are now part of the collection of Naturalis Biodiversity Center. The texts I used as the basis for the study of myrmecophily are Hölldobler & Wilson (1990) and Parker (2016). For the behavior of *Claviger*, I also used Cammaerts (1995, 1999). The great age of the *Claviger*–ant association and the total number of myrmecophiles are reported in Parker & Grimaldi (2014). The concept of ecosystem engineers is laid out in Jones *et al.* (1994). I obtained information on beavers as ecosystem engineers from Wright *et al.* (2002). The information on the Mannahatta Project comes from Reumer (2014), the project's main website, http://welikia.org, and also from Paumgarten (2007), Miller (2009), Sanderson & Brown (2007), Bean & Sanderson (2008) and Eric Sanderson's 2009 TED Talk (available via YouTube). Muhheakantuck is the original Lanape name for what today we call Hudson River.

CHAPTER 2: THE ANT(HROPO)-HILL

The trip to Maliau Basin Studies Center took place July 27–30, 2016. For more on this area and our work there, see www.taxonexpeditions.com. For the text about ecosystem engineering in hunter-gatherers, I used Marlowe (2005) and Smith (2007). The human trophic level is discussed in Bonhommeau *et al.* (2013). For the history of urbanization, I used Gross (2016), Reba *et al.* (2016), Newitz (2013), Misra (2015, 2016), The Data Team (2015) and Vance Kite's TED animation at https://ed.ted.com/lessons/. The animation of Reba's data can be viewed on https://youtu.be/yKJYXujJ7sU.

CHAPTER 3: DOWNTOWN ECOLOGY

The walk through Singapore with Chan Sow-Yan took place on August 2, 2016. I used the following sources. General urban ecology texts: McDonnell & MacGregor-Fors (2016) and Schmid (1978). General Singaporean urban ecology: Ward (1968), Lok & Lee (2009), Davison (2007) and Davison *et al.* (2008). On relics of original habitat in Singapore: Brook *et al.* (2003), Clements *et al.* (2005) and Lok *et al.* (2013). On organisms from rocky habitats utilizing buildings and walls: Ward (1968), Sipman (2009) and Tan *et al.* (2014). On the Singapore urban heat island: Chow & Roth (2006) and Roth & Chow (2012). On pollutants in Singapore: Xu *et al.* (2011), Sin *et al.* (2016) and Rothwell & Lee (2010). Use of human food by urban fauna in Singapore: Soh *et al.* (2002). On the spread of exotic species in Singapore: Tan & Yeo (2009), Chong *et al.* (2012), Ng & Tan (2010) and Teo *et al.* (2011). The collapse of Singaporean food webs: Jeevanandam & Corlett (2013). The exotic species in the San Francisco Bay area are mentioned in Cohen & Carlton (1998); see also Schilthuizen (2008). For the theory of island biogeography, and the Bracknell roundabout study, see MacArthur & Wilson (1967), Helden & Leather (2004), and Schilthuizen (2008). The introduced millipede is reported for Singapore in Decker & Tertilt (2012). The entire text about Singapore was checked by Chan Sow-Yan and Tan Siong-Kiat.

CHAPTER 4: URBAN NATURALISTS

For the story on the Rotterdam house crows, I used Nyári *et al.* (2006), De Baerdemaeker & Klaassen (2012), Hendriks (2014) and Dooren (2016). My visits to the live and deceased parts of the Hoek van Holland house crow population took place on August 17, 2016, as did my visit to the permanent exhibition of the Rotterdam Natural History Museum. More on the red squirrels in Rotterdam can be found in Moeliker (2015). Kees Moeliker checked and corrected my text on the Rotterdam museum. The *Mannahatta* book is Sanderson (2009). Information on Herbert Sukopp is from Reumer (2014). Further notes on the growth of urban ecology as a field are from Schilthuizen (2016b). The growth of citizen science as a means of biodiversity discovery is highlighted in Nielsen (2012). The Malaise trap results from Leicester are in Owen (1978). An introduction to the railway triangle inventories by the KNNV in Rotterdam is in Werf (1982). BioBlitzes are treated in detail in Baker *et al.* (2014). More about the City Nature Challenge 2017 can be found at http://www.calacademy.org/citizen-science /city-nature-challenge, and on *Belles de Bitume* at https://www.frederique -soulard-contes.com/belles-de-bitume. The BioBlitz in Wellington that resulted in a new diatom is in Harper *et al.* (2009). The new frog and snail species discovered in New York and São Paulo are in Feinberg *et al.* (2014) and Martins & Simone (2014), respectively. The new subterranean water beetles from cities in Japan are described in Uéno (1995). The numbers of beetle species from the Amsterdamse Bos are in Nonnekens (1961, 1965), and the flora of Brussels is mentioned in Godefroid (2001). A meta-analysis of 105 rural-to-urban gradient studies is in McKinney (2008).

CHAPTER 5: CITY SLICKERS

The reference to Schouw (1823) is in Sukopp (2008). The "luxury effect" is described in Hope *et al.* (2003). I found information on why urban centers are located in pre-existing biodiversity hotspots in Secretariat of the Convention on Biological Diversity (2012). The study from the Czech Republic that I mention is in Chocholoušková & Pyšek (2003). Higher biodiversity because of a greater variety of (micro)habitats is reviewed in Kowarik (2011). For the paragraphs on large vertebrates, I used Vyas (2012), Hoh (2016), Bateman & Fleming (2012), Soniak (2014), Mahoney (2012), Gehrt (2007), Jones (2009), and Baggaley (2014). The quote by Gehrt was taken from Mahoney (2012). The homepage of the BUGS project of the University of Sheffield is at http://www.bugs.group.shef.ac.uk. Later, when other cities were also included, the project was renamed Biodiversity of Urban GardenS. Here, I mainly used the following papers resulting from this project: Gaston *et al.* (2005), Smith *et al.* (2006a, 2006b). Kevin Gaston proofread the section on the BUGS project. Results similar to those of the BUGS project were obtained for Bangalore (Jaganmohan *et al.*, 2013) and Berlin (Zerke, 2003).

CHAPTER 6: IF I CAN MAKE IT THERE

The walk with Geerat Vermeij took place on June 17, 2014, whereas his quotes on pre-adaptation are from an email exchange with him in late September 2016. He also proofread and approved the text. Information on the house sparrow was taken from Anderson (2006). I used SOVON Vogelonderzoek Nederland (2012) for some details on natural and urban habitats of Dutch birds. Details on domestic arthropods are found in Bertone *et al.* (2016). The work on birds from Chilean cities is in Silva *et al.* (2016), and Carmen Paz and Olga Barbosa proofread this for me. The study on birds pre-adapted to human-generated noise is Francis *et al.* (2011), and Clint Francis proofread this section; I also used Woodsen (2011). The final paragraphs are based on Parker (2016), who also proofread this section for me.

CHAPTER 7: THESE ARE THE FACTS

The story on Albert Farn is based on Hart *et al.* (2010), Jenkinson (1922), Salmon *et al.* (2000), the website http://butterflyzoo.co.uk/farnfestival.html, and email correspondences with Adam Hart in June 2016, Stephen Sutton in June and October 2016, and Erik van Nieukerken in October 2016. The letter from Farn is under registration DCP-LETT-11747 at www.darwinproject .ac.uk (Darwin Correspondence Project, 2017). The section of this chapter on Farn was proofread by Adam Hart. Regarding Darwin's assessment of the speed of evolution, Hooper (2002: 55), citing Mayr & Provine (1980), writes that Darwin's son, Leonard Darwin, recalls that his father thought that it would take at least fifty generations for a change due to natural selection to be observable, suggesting that he could imagine evolution taking place within a human lifetime after all. For the various versions of *On the Origin of Species*, I consulted van Wyhe (2002). The online simulation of natural selection I did can be performed via http://www.radford.edu/~rsheehy /Gen_flash/popgen/.

CHAPTER 8: URBAN MYTHS

For the history of the peppered moth case, I used Cook (2003), Cook & Saccheri (2013), White (1877), Tutt (1896: 305–307), Haldane (1924), Cook *et al.* (1970, 1986, 2012), Kettlewell (1955, 1956), Coyne (1998), Hooper (2002), Van 't Hof *et al.* (2016), Rudge (2005), and Majerus (1998, 2009). The quote by Saccheri is from the *Nature* podcast at http://www.nature.com /nature/journal/v534/n7605/abs/nature17951.html. Laurence Cook read the whole chapter and helped me with some of the finer points.

CHAPTER 9: SO IT REALLY IS

The information on natural melanism in moths comes from Kettlewell (1973) and from email correspondence with Stephen Sutton on October 28, and November 3, 2016. Stephen Sutton also proofread those paragraphs. The biomechanics of bird wing shape is in Swaddle & Lockwood (2003). Wing shape evolution in starlings is based on Bitton & Graham (2015) and in cliff swallows on Brown & Bomberger-Brown (2013). Mary Bomberger-Brown read and corrected my text about the cliff swallow work. In an email to me of December 12, 2016, Pierre-Paul Bitton (who kindly checked my text about his work) expresses doubt that traffic, rather than pets, was behind the starling wing shape evolution. The work on *Crepis sancta* in Montpellier is described in Cheptou *et al.* (2008), while similar evolution on islands is in Cody & Overton (1996). Pierre-Olivier Cheptou kindly proofread my text about his work. My coverage of the work on rapid evolution in Caribbean *Anolis* is based on Losos *et al.* (1997), Marnocha *et al.* (2011), Tyler *et al.* (2016), and Winchell *et al.* (2016). As a source for *Anolis* biology in general, I used Losos (2009). Gingerich (1993) is the source for the rule-of-thumb calculation of the darwin unit. Kirstin Winchell read through my *Anolis* text and provided many useful comments.

CHAPTER 10: TOWN MOUSE, COUNTRY MOUSE

I viewed the ring-necked parakeets in Paris on December 15–17, 2016. The story that Jimi Hendrix released parakeets in London can be found in Brennan (2016). Human phylogeography is outlined in Harcourt (2016). The quotes by Ariane Le Gros are from email correspondence on December 14, 2016 and May 17, 2017. My sources for generalities on the species, including its invasion history and competition with nuthatches, are Strubbe & Matthysen (2009), Strubbe *et al.* (2010), Le Gros *et al.* (2016), the species page on the IUCN Red List (http://www.iucnredlist.org/details/22685441/0), its Wikipedia page https://en.wikipedia.org/wiki/Rose-ringed_parakeet, and the materials available via the homepage of ParrotNet (https://www.kent.ac.uk /parrotnet/). The relevant texts were read and approved by Ariane Le Gros. The information on bobcats comes from Serieys *et al.* (2015) and Riley *et al.* (2007), and the text was checked by Laurel Serieys. The information on global habitat fragmentation by roads is from Ibisch *et al.* (2016). My interview with Jason Munshi-South took place on December 10, 2016, and I also took some quotes from his TED-Ed lesson at http://ed.ted.com/lessons /evolution-in-a-big-city. His white-footed mouse publications that I used are: Munshi-South & Kharchenko (2010), Munshi-South & Nagy (2013),

Harris & Munshi-South (2013, 2016) and Harris *et al.* (2016). Jason Munshi-South read and checked my text about his work. The spider work is in Schäfer *et al.* (2001). A recent article shows fragmentation and adaptation in lizards in city parks in Brisbane (Littleford-Colquhoun *et al.*, 2017).

CHAPTER 11: POISONING PIGEONS IN THE PARK

The work on pollution-tolerant killifish is based on Whitehead *et al.* (2010, 2011, 2016) and Reid *et al.* (2016). I also used Kaplan (2016) and Carson (1962). The effects of PCBs and PAHs on AHR are based on the website of the Hahn lab at Woods Hole Oceanographic Institution (http://www.whoi .edu/science/B/people/mhahn/). Andrew Whitehead kindly proofread and corrected the section on the mummichog work; the quote at the end of this section is from an email of May 25, 2017. The section on de-icing salt is based on Coldsnow *et al.* (2017) and Houska (2016), whereas the principles of salt stress are from Mäser *et al.* (2002). Kayla Coldsnow kindly checked the paragraph I wrote about her work. Copper tolerance in monkey flowers is covered on pages 160–3 of my previous book, *Frogs, Flies and Dandelions* (Schilthuizen, 2001) and also in Wright *et al.* (2015). The zinc tolerance in grasses is from Al-Hiyali *et al.* (1990), and the work on melanic urban pigeons from Obhukova (2007) and Chatelain *et al.* (2014, 2016). Marion Chatelain proofread the section I wrote about her work.

CHAPTER 12: BRIGHT LIGHTS, BIG CITY

The attraction of birds by the 9/11 memorial lights was highlighted in a lecture by Kamiel Spoelstra in Haren, the Netherlands, on August 27, 2016. I also used the information on http://www.audubon.org/news/making -911-memorial-lights-bird-safe and https://www.sott.net/article/266370 -Thousands-of-migrating-birds-attracted-to-9-11-memorial-lights. The quotes by Michael Ahern were taken from a video at https://www.youtube.com /watch?v=LKPkJ08CBdc. The information on Euro 2016 and Silver-Y migration is from www.uefa.com, Moeliker (2016) and Chapman *et al.* (2012, 2013). For general information on light pollution I used a discussion on ResearchGate entitled, "Is there any convincing explanation yet about why moths are attracted to artificial light," Longcore & Rich (2004), Gaston *et al.* (2014) and a lecture by Kevin Gaston at Leiden University on June 30, 2016. Kevin Gaston also proofread part of this section. The anecdotal data on bird deaths due to light come from Guynup (2003) and those on insect deaths from Eisenbeis (2006). The example of birds killed at a lighthouse is from Jones & Francis (2003). A list of International Dark-Sky

Parks is at http://darksky.org/idsp/parks/. The papers on adaptation to light in moths and spiders, respectively, are Altermatt & Ebert (2016) and Heiling (1999). I used information from an email correspondence with Florian Altermatt in February 2017, and he also proofread the text I wrote on his work.

CHAPTER 13: BUT IS IT REALLY EVOLUTION?

The creationist blog post that I mention is at http://darwins-god.blogspot.nl /2017/01/evolutionist-evolution-is-happening.html. The difference between soft and hard selection is explained in, for example, Wright (2015) and Hermisson & Pennings (2017). For the details on the selection in mummichog and peppered moth, see the notes to Chapters 11 and 8, respectively. For general considerations on learning and epigenetics, I used Skinner (2011), Azzi *et al.* (2014) and Arney (2017). Plasticity in ground hopper color is in Hochkirch *et al.* (2008). The quote by Kevin Gaston is from Evans *et al.* (2010). As I was editing the final version of this book, the first article appeared that showed that epigenetics causes important differences between urban and rural Darwin's finches (McNew *et al.*, 2017).

CHAPTER 14: CLOSE URBAN ENCOUNTERS

The work on pigeon predation by catfish is in Cucherousset *et al.* (2012) and discussed in Yong (2012). I also used an email correspondence with Cucherousset and Santoul of March, 2017. Frédéric Santoul kindly proofread my text about the catfish work. The paper on acorn ants is Diamond *et al.* (2017), and I also used the blog post "A tale of two thousand cities" on Andrew Hendry's blog ecoevoevoeco.blogspot.com. The food eaten by ring-necked parakeets in Paris is from Clergeau *et al.* (2009). The pigeons eating hibiscus buds is something I observed myself in Kota Kinabalu. The work on soapberry bugs is described in, for example, Carroll *et al.* (2001, 2005). Scott Carroll kindly checked my text on his work. In my earlier book *Frogs, Flies, and Dandelions* (Schilthuizen, 2001), I tell the story of the apple maggot fly. For the corn borer work, see Calcagno *et al.* (2010). The beetle that shifted to American black cherry is introduced in Schilthuizen *et al.* (2016). Daehler & Strong (1997) report on the defense by introduced cordgrass. Curtis Daehler kindly proofread this section for me and points out that *Prokelisia marginata* has been introduced to Willapa Bay after their study was completed. The birds that use cigarette butts in their nests are described in Suárez-Rodríguez *et al.* (2013), and I also used information from http://www.cigwaste.org. This bit of text was checked by Isabel López-Rull.

CHAPTER 15: SELF-DOMESTICATION

The original publications on the Sendai crows are Nihei (1995) and Nihei & Higuchi (2001). I thank Satoshi Chiba, Osamu Mikami, Minoru Chiba, Yawara Takeda, and my local nut grocer for their help in our (unsuccessful) attempts to see the behavior in Sendai, Japan, in May and June 2017. The behavior was filmed for episode 10 in the 1998 David Attenborough BBC TV series *The Life of Birds* (see https://www.youtube.com/watch?v=BGPGknpq3e0). For the account on milk bottle-opening tits, I used Fisher & Hinde (1949, 1951) and Lefebvre (1995). The record of fifty-seven bottles opened at a school was from Cramp *et al.* (1960). For the work on tits' problem-solving abilities and social transmission, Aplin *et al.* (2013, 2015). I also used the video at http://www.dailymail.co.uk/sciencetech/article-2868613/Great-tits -pass-traditions-adapt-fit-locals.html. The text on Barbados bullfinch is based on Audet *et al.* (2016), his November 9, 2016, blog on http://ecoevoevoeco .blogspot.com/2016/11/street-smarts.html, and an email correspondence with him on April 29, 2017. Jean-Nicolas Audet also read and approved my text about his work. The work on neophilia in Polish cities is Tryjanowski *et al.* (2016), and Piotr Tryjanowksi also kindly proofread the text. The other examples of neophobia work in chickadees, crows, and mynahs is from Williams (2009), Greggor (2016) and Sol *et al.* (2011), respectively. The tolerance study is Symonds *et al.* (2016). Matt Symonds proofread the paragraph about his work.

CHAPTER 16: SONGS OF THE CITY

The "urban acoustic ecology" practical that I describe took place on September 9, 2016. For general acoustic ecology, I made use of Warren *et al.* (2006) and Swaddle *et al.* (2015). Slabbekoorn's work on great tits was first described in Slabbekoorn & Peet (2003). Similar work under natural conditions was published by Hunter & Krebs (1979). The examples of other urban birds changing their song are from Slabbekoorn (2013) and of treefrogs from Parris *et al.* (2009). The grasshopper work is Lampe *et al.* (2012, 2014), and the chiffchaff study is Verzijden *et al.* (2010). The work on songbirds' and non-songbirds' calls and songs are Hu & Cardoso (2010) and Potvin *et al.* (2011). The studies on the effects of changed great tits' songs on rival males and females are Mockford & Marshall (2009) and Halfwerk *et al.* (2011), respectively. Millie Mockford and Wouter Halfwerk proofread the paragraphs about their respective work. The other "kinds" of urban acoustic ecology are: more "hurried" silvereye songs (Potvin *et al.*, 2011), night-time singing in robins (Fuller *et al.*, 2007), and birds advancing their song near airports (Gil *et al.*, 2015). This entire chapter was checked by Hans Slabbekoorn.

CHAPTER 17: SEX AND THE CITY

For the part on the dark-eyed junco, I mainly used Yeh (2004), Shochat *et al.* (2006), McGlothlin *et al.* (2008) and Hill *et al.* (1999). I also used an email correspondence with Pamela Yeh of May 2017. Pamela Yeh proofread the section about her work. For the story about the Barcelona great tits, I used Galván & Alonso-Alvarez (2008), Senar *et al.* (2014), and Bjørklund *et al.* (2010). Juan Carlos Senar also checked my text about this work. The damselfly work is Tüzün *et al.* (2017), and my text was checked by Nedim Tüzün and Lin Op de Beeck. The work with the robo-squirrel is in Partan *et al.* (2010), the information on Indian gerbils comes from Hutton & McGraw (2016) and that on hormone-mimicking chemicals is based on Zala & Penn (2004). The two examples for Australian "evolutionary traps" are from Gwynne & Rentz (1983) and an email correspondence with Cat Davidson, relayed to me by Bronwen Scott. For the bowerbirds, I also used https://www.zoo.org.au/news/feeling-blue. David Rentz proofread the section about beetles and beer bottles.

CHAPTER 18: TURDUS URBANICUS

Galápagos evolution is summarized by Parent *et al.* (2008). For more detail on the Darwin's finches, see my book *Frogs, Flies, and Dandelions* (Schilthuizen, 2001). Good overviews of the ongoing evolutionary research on Darwin's finches are Weiner (1995) and Hendry (2017). The work on the incipient splitting of *Geospiza fortis* on Santa Cruz island (and its breakdown near the city) is in Hendry *et al.* (2006) and De Léon *et al.* (2011, 2017), as well as several papers they cite. The earliest mention of the urban blackbird is Bonaparte (1827). My history of blackbird urbanization is largely based on Evans *et al.* (2010) and Møller *et al.* (2014). The differences in body shape are from Grégoire (2003), Lippens & Van Hengel (1962), and Evans *et al.* (2009a). Song pitch and timing were studied by, respectively, Ripmeester *et al.* (2010) and Nordt & Klenke (2013). The research on timing of reproduction is by Partecke *et al.* (2004). The migration study is Partecke & Gwinner (2007) and the stress hormone study is Partecke *et al.* (2006) and Müller *et al.* (2013). Jesko Partecke kindly checked the text I wrote about his work. The flight initiation distance differences are from Symonds *et al.* (2016). The DNA fingerprinting work is Evans *et al.* (2009b). I was first alerted to the urbanity of Darwin's finches by Barbara Waugh. The research mentioned is from De León (2011, 2017), who also proofread the relevant sections of this chapter.

CHAPTER 19: EVOLUTION IN A TELECOUPLED WORLD

For the history of Von Siebold and Japanese knotweed, I used their respective Wikipedia pages (accessed June 7, 2017), Christenhusz (2002), Christenhusz & van Uffelen (2001) and Peeters (2015). Norbert Peeters kindly proofread the Von Siebold text for me. Some further examples of globally transported species come from Thompson (2014) and Schmidt *et al.* (2017). The original coining of the term "supertramp" is Diamond (1974). The work on homogenization in microbes, birds and street flora is from Schmidt *et al.* (2017), Murthy *et al.* (2016), and Wittig & Becker (2010). Marina Alberti's ideas and quotes are from a Skype interview I had with her on September 8, 2016, and from her papers Alberti (2015) and Alberti *et al.* (2003, 2017). Marina Alberti also proofread the piece I wrote about her work and ideas. The work of Kamiel Spoelstra is at: https://nioo.knaw.nl/nl/employees/kamiel-spoelstra. The quote about intercity connectivity is from Khanna (2016). I publicized my own genome sequencing project via my blog and programs in 2015 on Dutch science radio show *De Kennis van Nu* (https://dekennisvannu.nl/site/special /De-code-van-Menno/8). Papers showing recent evolution in humans are Field *et al.* (2016) and Barnes *et al.* (2011). I also used Bolhuis *et al.* (2011), Pennisi (2016), and Hassell *et al.* (2016).

CHAPTER 20: DESIGN IT WITH DARWIN

Our visit to Roppongi Hills took place on May 29, 2017. A request by email to the press officer at Mori Building Corporation for more information on the company's green roofing policies was not answered. I took information on green roofs in Japan from https://resources.realestate.co.jp/living/japan-green -roof-buildings/, as well as the company websites of Mori Building Company and Emilio Ambasz & Associates. Information on Singapore's green walls comes from http://inhabitat.com/tag/green-skyscrapers/. I took information on rooftop farming from Hui *et al.* (2011). More about New York's Lowline can be found at http://thelowline.org, and on Berlin's "green mountain" at: http:// www.hilldegarden.org. My information on the Tokyo City ordinance on green roofs is from http://www.c40.org/case_studies/nature-conservation-ordinance -is-greening-tokyo-s-buildings. The books I mention are Vink *et al.* (2017), Gunnell *et al.* (2013), and Dunnet & Kingsbury (2004). A video with English subtitles on the work of *Gewildgroei* can be found at https://vimeo.com /175805142. More on the native/non-native debate is at the Global Roundtable at http://www.thenatureofcities.com for November 5, 2015. I also used Davis *et al.* (2011), Foster & Sandberg (2004) and Johnston *et al.* (2011). The examples of embedded primary forest come from Tan & Jim (2017) and Diogo *et al.* (2014). More on the corridor debate is at the Global Roundtable at http://www .thenatureofcities.com for October 5, 2014. My visit to Tokyo Metropolitan University took place on May 26, 2017. For the concept of Satoyama, I used Kobori & Primack (2003), Kohsaka *et al.* (2013) and Puntigam *et al.* (2010), as well as http://satoyama-initiative.org. For SnailSnap, see: http://snailsnap. nl. See Farina *et al.* (2014) and the website https://naturesmartcities.com/ for some background on soundscape citizen science. For the Funky Nest Contest, see: http://nestwatch.org. The three potential research questions on "funky" urban bird nests are inspired by Wang *et al.* (2015), Sergio *et al.* (2011) and Suárez-Rodríguez *et al.* (2013).

OUTSKIRT

Find out more about our Borneo expeditions at http://www.taxonexpeditions .com. The term "hyperkeystone species" comes from Worm & Paine (2016). My correspondence with Sabine Rietkerk took place on June 27, 2017.

BIBLIOGRAPHY

Al-Hiyaly, S.A.K., T. McNeilly & A.D. Bradshaw, 1990. The effect of zinc contamination from electricity pylons. Contrasting patterns of evolution in five grass species. *New Phytologist*, 114: 183–190.

Alberti, M., 2015. Eco-evolutionary dynamics in an urbanizing planet. *Trends in Ecology and Evolution*, 30: 114–126.

Alberti, M., J.M. Marzluff, E. Shulenberger, G. Bradley, C. Ryan & C. Zumbrunnen, 2003. Integrating humans into ecology: opportunities and challenges for studying urban ecosystems. *BioScience*, 53: 1169–1179.

Alberti, M., C. Correa, J.M. Marzluff, A.P. Hendry, E.P. Palkovacs, K.M. Gotanda, V.M. Hunt, T.M. Apgar & Y. Zhou, 2017. Global urban signatures of phenotypic change in animal and plant populations. *Proceedings of the National Academy of Sciences*, 201606034.

Altermatt, F. & D. Ebert, 2016. Reduced flight-to-light behavior of moth populations exposed to long-term urban light pollution. *Biology Letters*, 12: 20160111.

Anderson, T.R., 2006. *Biology of the Ubiquitous House-Sparrow: From Genes to Populations*. Oxford: Oxford University Press. xi+547 pp.

Aplin, L.M., D.R. Farine, J. Morand-Ferron, A. Cockburn, A. Thornton & B.C. Sheldon, 2015. Experimentally induced innovations lead to persistent culture via conformity in wild birds. *Nature*, 518: 538–541.

Aplin, L.M., B.C. Sheldon & J. Morand-Ferron, 2013. Milk bottles revisited: social learning and individual variation in the blue tit, *Cyanistes caeruleus*. *Animal Behavior*, 85: 1225–1232.

Arney, K., 2017. What is epigenetics? *Little Atoms*, 2.

Audet, J.N., S. Ducatez & L. Lefebvre, 2015. The town bird and the country bird: problem solving and immunocompetence vary with urbanization. *Behavioral Ecology*, 27: 637–644.

Azzi, A., R. Dallmann, A. Casserly, H. Rehrauer, A. Patrignani, B. Maier, A. Kramer & S.A. Brown, 2014. Circadian behavior is light-reprogrammed by plastic DNA methylation. *Nature Neuroscience*, 17: 377–382.

Baerdemaeker, A. de & O. Klaassen, 2012. Huiskraaien in Hoek van Holland: is de groei eruit? *Straatgras*, 24: 78–79.

Baggaley, K., 2014. Cities are brimming with wildlife worth studying. *Science-News*, December 29, 2014.

Baker, G.M., N. Duncan, T. Gostomski, M.A. Horner & D. Manski, 2014. The bioblitz: good science, good outreach, good fun. *Park Science*, 31: 39–45.

Barnes, I., A. Duda, O.G. Pybus & M.G. Thomas, 2011. Ancient urbanization predicts genetic resistance to tuberculosis. *Evolution*, 65: 842–848.

Bateman, P.W. & P.A. Fleming, 2012. Big city life: carnivores in urban environments. *Journal of Zoology*, 287: 1–23.

Bean, W.T. & E.W. Sanderson, 2008. Using a spatially explicit ecological model to test scenarios of fire use by Native Americans: An example from the Harlem Plains, New York, NY. *Ecological Modeling*, 211: 301–308.

Bertone, M.A., M. Leong, K.M. Bayless, T.L. Malow, R.R. Dunn & M.D. Trautwein, 2016. Arthropods of the great indoors: characterizing diversity inside urban and suburban homes. *PeerJ*, 4: e1582.

Bitton, P.-P. & B.A. Graham, 2015. Change in wing morphology of the European starling during and after colonization of North America. *Journal of Zoology*, 295: 254–260.

Bjørklund, M., I. Ruiz & J.C. Senar, 2010. Genetic differentiation in the urban habitat: the great tits (*Parus major*) of the parks of Barcelona city. *Biological Journal of the Linnean Society*, 99: 9–19.

Bolhuis, J.J., G.R. Brown, R.C. Richardson & K.N. Laland, 2011. Darwin in mind: New opportunities for evolutionary psychology. *PLoS Biology*, 9: e1001109.

Bonaparte, C.L., 1827. *Specchio Comparativo delle Ornitologie di Roma e de Filadelfia*. Pisa, 80 pp.

Bonhommeau, S., L. Dubroca, O. Le Pape, J. Barde, D.M. Kaplan, E. Chassot & A.E. Nieblas, 2013. Eating up the world's food web and the human trophic level. *Proceedings of the National Academy of Sciences*, 110: 20617–20620.

Brennan, A., 2016. Is Jimi Hendrix responsible for London's parakeet population? *GQ Magazine*, February 10, 2016. http://www.gq-magazine.co.uk/article/jimi-hendrix-parakeets-hampstead-heath-kingston-primrose-hill.

Brook, B.W., N.S. Sodhi & P.K.L. Ng, 2003. Catastrophic extinctions follow deforestation in Singapore. *Nature*, 424: 420–423.

Brown, C.R. & M. Bomberger-Brown, 2013. Where has all the road kill gone? *Current Biology*, 23: R233–R234.

Byrne, K. & R.A. Nichols, 1999. *Culex pipiens* in London Underground tunnels: differentiation between surface and subterranean populations. *Heredity*, 82: 7–15.

Calcagno, V., V. Bonhomme, Y. Thomas, M.C. Singer & D. Bourguet, 2010.

Divergence in behavior between the European corn borer, *Ostrinia nubilalis*, and its sibling species *Ostrinia scapulalis*: adaptation to human harvesting? *Proceedings of the Royal Society of London B: Biological Sciences*, rspb20100433.

Cammaerts, R., 1995. Regurgitation behavior of the *Lasius flavus* worker (Formicidae) toward the myrmecophilous beetle *Claviger testaceus* (Pselaphidae) and other recipients. *Behavioral Processes*, 34: 241–264.

Cammaerts, R., 1999. Transport location patterns of the guest beetle *Claviger testaceus* (Pselaphidae) and other objects moved by workers of the ant, *Lasius flavus* (Formicidae). *Sociobiology*, 34: 433–475.

Carroll, S.P., H. Dingle, T.R. Famula & C.W. Fox, 2001. Genetic architecture of adaptive differentiation in evolving host races of the soapberry bug, *Jadera haematoloma*. *Genetica*, 112: 257–272.

Carroll, S.P., J.E. Loye, H. Dingle, M. Mathieson, T.R. Famula & M. Zalucki, 2005. And the beak shall inherit—evolution in response to invasion. *Ecology Letters*, 8: 944–951.

Carson, R., 1962. *Silent Spring.* New York: Houghton Mifflin.

Chapman, J.W., J.R. Bell, L.E. Burgin, D.R. Reynolds, L.B. Pettersson, J.K. Hill, M.B. Bonsall & J.A. Thomas, 2012. Seasonal migration to high latitudes results in major reproductive benefits in an insect. *Proceedings of the National Academy of Sciences*, 109: 14924–14929.

Chapman, J.W., K.S. Lim & D.R. Reynolds, 2013. The significance of midsummer movements of *Autographa gamma*: Implications for a mechanistic understanding of orientation behavior in a migrant moth. *Current Zoology*, 59: 360–370.

Chatelain, M., J. Gasparini & A. Frantz, 2016. Do trace metals select for darker birds in urban areas? An experimental exposure to lead and zinc. *Global Change Biology*, 22: 2380–2391.

Chatelain, M., J. Gasparini, L. Jacquin & A. Frantz, 2014. The adaptive function of melanin-based plumage coloration to trace metals. *Biology Letters*, 10: 20140164.

Cheptou, P.-O., O. Carrue, S. Rouifed & A. Cantarel, 2008. Rapid evolution of seed dispersal in an urban environment in the weed *Crepis sancta*. *Proceedings of the National Academy of Science USA*, 105: 3796–3799.

Chong, K.Y., S. Teo, B. Kurukulasuriya, Y.F. Chung, S. Rajathurai, H. C. Lim & H.T.W. Tan, 2012. Decadal changes in urban bird abundance in Singapore. *Raffles Bulletin of Zoology*, S25: 189–196.

Chocholoušková, Z. & P. Pyšek, 2003. Changes in composition and structure of urban flora over 120 years: a case study of the city of Plzen. *Flora*, 198: 366–376.

Chow, W.T.L. & M. Roth, 2006. Temporal dynamics of the urban heat island of Singapore. *International Journal of Climatology*, 26: 2243–2260.

Christenhusz, M.J.M., 2002. Planthunter Von Siebold. *Dirk van der Werffs Plants*, 7 (1): 36–38.

Christenhusz, M.J.M. & G.A. van Uffelen, 2001. Verwilderde Japanse planten in Nederland, ingevoerd door Von Siebold. *Gorteria*, 27: 97–108.

Clements, R., L.P. Koh, T.M. Lee, R. Meyer & D. Li, 2005. Importance of reservoirs for the conservation of freshwater molluscs in a tropical urban landscape. *Biological Conservation*, 128: 136–146.

Clergeau, P., A. Vergnes & R. deLanque, 2009. La perruche à collier *Psittacula krameri* introduite en Île-De-France: distribution et régime alimentaire. *Alauda* 77: 121–132.

Cody, M.L. & J.M. Overton, 1996. Short-term evolution of reduced dispersal in island plant populations. *Journal of Ecology*, 84: 53–61.

Cohen, A.N. & J.T. Carlton, 1998. Accelerated invasion rate in a highly invaded estuary. *Science*, 279: 555–558.

Coldsnow, K.D., B.M. Mattes, W.D. Hintz & R.A. Relyea, 2017. Rapid evolution of tolerance to road salt in zooplankton. *Environmental Pollution*, 222: 367–373.

Cook, L.M., 2003. The rise and fall of the *carbonaria* form of the peppered moth. *The Quarterly Review of Biology*, 78: 399–417.

Cook, L.M., R.R. Askew & J.A. Bishop, 1970. Increased frequency of the typical form of the peppered moth in Manchester. *Nature*, 227: 1155.

Cook, L.M., B.S. Grant, I.J. Saccheri & J. Mallet, 2012. Selective bird predation on the peppered moth: the last experiment of Michael Majerus. *Biology Letters*, 8: 609–612.

Cook, L.M., G.S. Mani & M.E. Varley, 1986. Postindustrial melanism in the peppered moth. *Science*, 231: 611–613.

Cook, L.M. & I.J. Saccheri, 2013. The peppered moth and industrial melanism: evolution of a natural selection case study. *Heredity*, 110: 207–212.

Coyne, J. A., 1998. Not black and white. Review of "melanism: evolution in action" by Michael E.N. Majerus. *Nature*, 396: 35–36.

Cramp, S., A. Pettet & J.T.R. Sharrock, 1960. The irruption of tits in autumn 1957. *British Birds*, 53: 49–77.

Cucherousset, J., S. Boulêtreau, F. Azémar, A. Compin, M. Guillaume & F. Santoul, 2012. "Freshwater Killer Whales:" Beaching behavior of an alien fish to hunt land birds. *PLoS ONE*, 7: e50840.

Daehler, C.C. & D.R. Strong, 1997. Reduced herbivore resistance in introduced smooth cordgrass (*Spartina alterniflora*) after a century of herbivore-free growth. *Oecologia*, 110: 99–108.

Darwin Correspondence Project, 2017. "Letter no. 11747," accessed on June 9, 2017, http://www.darwinproject.ac.uk/DCP-LETT-11747.

Davis, M.A., M.K. Chew, R.J. Hobbs, A.E. Lugo, J.J. Ewel, G.J. Vermeij, J.H. Brown, M.L. Rosenzweig, M.R. Gardner, S.P. Carroll, K. Thompson, S.T.A. Pickett, J.C. Stromberg, P. Del Tredici, K.N. Suding, J.G. Ehrenfeld, J.P. Grime, J. Mascaro & J.C. Briggs, 2011. Don't judge species on their origins. *Nature*, 474: 153–154.

Davison, G.W.H., 2007. Urban forest rehabilitation—a case study from Singapore. Pp. 171–181 in: (D.K. Lee, ed.) *Keep Asia Green; Vol. 1: "Southeast Asia."* IUFRO, Vienna, Austria.

Davison, G.W.H., P.K.L. Ng & H.C. Ho, 2008. *The Singapore Red Data Book: Threatened Plants and Animals of Singapore*. 2nd edition. Nature Society (Singapore), Singapore. 285 pp.

Decker, P. & T. Tertilt, 2012. First records of two introduced millipedes *Anoplodesmus saussurii* and *Chondromorpha xanthotricha* (Diplopoda: Polydesmida: Paradoxosomatidae) in Singapore. *Nature in Singapore*, 5: 141–149.

De León, L.F., J.A. Raeymaekers, E. Bermingham, J. Podos, A. Herrel & A.P. Hendry, 2011. Exploring possible human influences on the evolution of Darwin's finches. *Evolution*, 65: 2258–2272.

De León, L.F., D.M.T. Sharpe, K.M. Gotanda, J.A.M. Raeymaekers, J.A. Chaves, A.P. Hendry & J. Podos, 2017. Human foods erode niche segregation in Darwin's finches. *Evolutionary Applications*. (in press).

Diamond, J.M., 1974. Colonization of exploded volcanic islands by birds: the supertramp strategy. *Science*, 184: 803–806.

Diamond, S.E., L. Chick, A. Perez, S.A. Strickler & R.A. Martin, 2017. Rapid evolution of ant thermal tolerance across an urban-rural temperature cline. *Biological Journal of the Linnean Society* doi: 10.1093/biolinnean/blw047.

Diogo, I.J.S., A.E.R. Holanda, A.L. de Oliveira Filho & C.L.F. Bezerra, 2014. Floristic composition and structure of an urban forest remnant of Fortaleza, Ceará. *Gaia Scientia*, 8: 266–278.

Donihue, C.M. & M.R. Lambert, 2015. Adaptive evolution in urban ecosystems. *AMBIO*, 3: 194–203.

Dooren, T. van, 2016. The unwelcome crows. *Angelaki*, 21: 193–212.

Dunnett, N. & N. Kingsbury, 2004. *Planting green roofs and living walls*. Portland, OR: Timber Press.

Eisenbeis, G., 2006. Artificial night lighting and insects: Attraction of insects to streetlamps in a rural setting in Germany. Pp. 281–304 in: *Ecological Consequences of Artificial Night Lighting* (C. Rich & T. Longcore, eds.). Washington, D.C.: Island Press.

Elfferich, C., 2011. *Natuur Dichtbij; Gewone en Ongewone Natuur in Pijnacker.* Caroline Elferrich, Pijnacker, the Netherlands. 84 pp.

Evans, K.L., K.J. Gaston, S.P. Sharp, A. McGowan & B.J. Hatchwell, 2009a. The effect of urbanisation on avian morphology and latitudinal gradients in body size. *Oikos*, 118: 251–259.

Evans, K.L., K.J. Gaston, A.C. Frantz, M. Simeoni, S.P. Sharp, A. McGowan, D.A. Dawson, K. Walasz, J. Partecke, T. Burke & B.J. Hatchwell, 2009b. Independent colonization of multiple urban centers by a formerly forest specialist bird species. *Proceedings of the Royal Society of London B*: rspb.2008.1712.

Evans, K.L., B.J. Hatchwell, M. Parnell & K.J. Gaston, 2010. A conceptual framework for the colonization of urban areas: the blackbird *Turdus merula* as a case study. *Biological Reviews*, 85: 643–667.

Farina, A., P. James, C. Bobryk, N. Pieretti, E. Lattanzi & J. McWilliam, 2014. Low cost (audio) recording (LCR) for advancing soundscape ecology toward the conservation of sonic complexity and biodiversity in natural and urban landscapes. *Urban ecosystems*, 17: 923–944.

Feinberg, J.A., C.E. Newman, G.J. Watkins-Colwell, M.D. Schlesinger, B. Zarate, *et al.*, 2014. Cryptic diversity in metropolis: Confirmation of a new leopard frog species (Anura: Ranidae) from New York City and surrounding Atlantic coast regions. *PLoS ONE*, 9: e108213.

Field, Y., E.A. Boyle, N. Telis, Z. Gao, K.J. Gaulton, D. Golan, L. Yengo, G. Rocheleau, P. Froguel, M.I. McCarthy & J.K. Pritchard, 2016. Detection of human adaptation during the past 2000 years. *Science*, 354: 760–764.

Fisher, J. & R.A. Hinde, 1949. The opening of milk bottles by birds. *British Birds*, 42: 347–357.

Fonseca, D.M., N. Keyghobadi, C.A. Malcolm, C. Mehmet, F. Schaffner, M. Mogi, R.C. Fleischer & R.C. Wilkerson, 2004. Emerging vectors in the *Culex pipiens* complex. *Science*, 303: 1535–1538.

Foster, J. & L.A. Sandberg, 2004. Friends or foe? Invasive species and public green space in Toronto. *Geographical Review*, 94: 178–198.

Francis, C.D., C.P. Ortega & A. Cruz, 2011. Noise pollution filters bird communities based on vocal frequency. *PLoS ONE*, 6: e27052.

Fuller, R.A., P.H. Warren & K.J. Gaston, 2007. Daytime noise predicts nocturnal singing in urban robins. *Biology Letters*, 3: 368–370.

Galván, I. & C. Alonso-Alvarez, 2008. An intracellular antioxidant determines the expression of a melanin-based signal in a bird. *PLoS ONE*, 3: e3335.

Gaston, K.J., P.H. Warren, K. Thompson & R.M. Smith, 2005. Urban domestic gardens (IV): the extent of the resource and its associated features. *Biodiversity and Conservation*, 14: 3327–3349.

Gaston, K.J., J.P. Duffy, S. Gaston, J. Bennie & T.W. Davies, 2014. Human alteration of natural light cycles: causes and ecological consequences. *Oecologia*, 176: 917–931.

Gehrt, S.D., 2007. Ecology of coyotes in urban landscapes. *Wildlife Damage Management Conferences—Proceedings*: Paper 63.

Gil, D., M. Honarmand, J. Pascual, E. Pérez-Mena & C. Macías Garcia, 2015. Birds living near airports advance their dawn chorus and reduce overlap with aircraft noise. *Behavioral Ecology*, 26: 435–443.

Gingerich, P.D., 1993. Quantification and comparison of evolutionary rates. *American Journal of Science*, 293A: 453–478.

Godefroid, S., 2001. Temporal analysis of the Brussels flora as indicator for changing environmental quality. *Landscape and Urban Planning*, 52: 203–224.

Greggor, A.L., N.S. Clayton, A.J. Fulford & A. Thornton, 2016. Street smart: faster approach toward litter in urban areas by highly neophobic corvids and less fearful birds. *Animal Behavior*, 117: 123–133.

Grégoire, A., 2003. *Démographie et différenciation chez le Merle noir Turdus merula: liens avec l'habitat et les relations hôtes-parasites.* Doctoral dissertation, Dijon.

Gross, M., 2016. The urbanisation of our species. *Current Biology*, 26: R1205-R1208.

Gunnell, K., C. Williams & B. Murphy, 2013. *Designing for Biodiversity: A Technical Guide for New and Existing Buildings.* RIBA Publishing, London.

Guynup, S., 2003. Light pollution taking toll on wildlife, eco-groups say. *National Geographic Today*, April 17, 2003.

Gwynne, D.T. & D.C.F. Rentz, 1983. Beetles on the bottle: male buprestids mistake stubbies for females (Coleoptera). *Journal of the Australian Entomological Society*, 22: 79–80.

Haberl, H., K.H. Erb, F. Krausmann, V. Gaube, A. Bondeau, C. Plutzar, S. Gingrich, W. Lucht & M. Fischer-Kowalski, 2007. Quantifying and mapping the human appropriation of net primary production in earth's terrestrial ecosystems. *Proceedings of the National Academy of Sciences of the USA*, 104: 12942–12947.

Haldane, J.B.S., 1924. A mathematical theory of natural and artificial selection. *Transactions of the Cambridge Philosophical Society*, 23: 19–41.

Halfwerk, W., S. Bot, J. Buikx, M. van der Velde, J. Komdeur, C. ten Cate & H. Slabbekoorn, 2011. Low-frequency songs lose their potency in noisy urban conditions. *Proceedings of the National Academy of Sciences USA*, 108: 14549–14554.

Harcourt, A.H., 2016. Human phylogeography and diversity. *Proceedings of the National Academy of Sciences of the USA*, 113: 8072–8078.

Harper, M.A., D.G. Mann & J.E. Patterson, 2009. Two unusual diatoms from New Zealand: *Tabularia variostriata* a new species and *Eunophora berggrenii*. *Diatom Research*, 24: 291–306.

Hart, A.G., R. Stafford, A.L. Smith & A.E. Goodenough, 2010. Evidence for contemporary evolution during Darwin's lifetime. *Current Biology*, 20: R95.

Hassell, J.M., M. Begon, M.J. Ward & E.M. Fèvre, 2017. Urbanization and disease emergence: dynamics at the wildlife–livestock–human interface. *Trends in Ecology & Evolution*, 32: 55–67.

Heiling, A.M., 1999. Why do nocturnal orb-web spiders (Araneidae) search for light? *Behavioral Ecology and Sociobiology*, 46: 43–49.

Helden, A.J. & S.R. Leather, 2004. Biodiversity on urban roundabouts—Hemiptera, management and the species-area relationship. *Basic and Applied Ecology*, 5: 367–377.

Hendriks, D., 2014. Woede in Hoek van Holland om afschieten huiskraaien. *Algemeen Dagblad*, March 6, 2014.

Hendry, A.P., 2017. *Eco-Evolutionary Dynamics*. Princeton: Princeton University Press, 416 pp.

Hendry, A.P., P.R. Grant, B.R. Grant, H.A. Ford, M.J. Brewer & J. Podos, 2006. Possible human impacts on adaptive radiation: beak size bimodality in Darwin's finches. *Proceedings of the Royal Society of London B*, 273: 1887–1894.

Hermisson, J. & P.S. Pennings, 2017. Soft sweeps and beyond: Understanding the patterns and probabilities of selection footprints under rapid adaptation. *BioRxiv*, doi: http://dx.doi.org/10.1101/114587.

Hill, J.A., D.A. Enstrom, E.D. Ketterson, V. Nolan & C. Ziegenfus, 1999. Mate choice based on static versus dynamic secondary sexual traits in the dark-eyed junco. *Behavioral Ecology*, 10: 91–96.

Hinde, R.A. & J. Fisher, 1951. Further observations on the opening of milk bottles by birds. *British Birds*, 44: 393–396.

Hochkirch, A., J. Deppermann & J. Gröning, 2008. Phenotypic plasticity in insects: the effects of substrate color on the coloration of two ground-hopper species. *Evolution & Development*, 10: 350–359.

Hof, A.E. van 't, P. Campagne, D.J. Rigden, C.J. Yung, J. Lingley, M.A. Quail, N. Hall, A.C. Darby & I.J. Saccheri, 2016. The industrial melanism mutation in British peppered moths is a transposable element. *Nature*, 534: 102–105.

Hoh, A., 2016. Brush turkeys invading suburban Sydney backyards. *ABC News*, March 31, 2016.

Hölldobler, B. & E.O. Wilson, 1990. *The Ants*. Belknap Press, Cambridge, Massachussetts, USA.

Hooper, J., 2002. *Of Moths and Men. Intrigue, Tragedy and the Peppered Moth.* New York: Fourth Estate, 400 pp.

Hope, D., C. Gries, W. Zhu, W.F. Fagan, C.L. Redman, N.B. Grimm, A.L. Nelson, C. Martin & A. Kinzig, 2003. Socioeconomics drive urban plant diversity. *Proceedings of the National Academy of Sciences USA*, 10: 8788–8792.

Houska, C., 2016. Deicing salt—recognizing the corrosion threat. http://www.imoa.info/download_files/stainless-steel/DeicingSalt.pdf

Hu, Y. & G.C. Cardoso, 2010. Which birds adjust the frequency of vocalizations in urban noise? *Animal Behavior*, 79: 863–867.

Hui, S.C.M., 2011. Green roof urban farming for buildings in high-density urban cities. The 2011 Hainan China World Green Roof Conference, March 18–21, 2011, 9 pp.

Huisman, J. & M. Schilthuizen, 2010. Vinex-merel is andere vogel dan z'n voorvader. *De Volkskrant*, November 15, 2010.

Hunter, L.M. & J.R. Krebs, 1979. Geographical variation in the song of the great tit (*Parus major*) in relation to ecological factors. *Journal of Animal Ecology*, 48: 759–785.

Hutton, P. & K.J. McGraw, 2016. Urban impacts on oxidative balance and animal signals. *Frontiers in Ecology and Evolution*, 4: 54.

Ibisch, P.L., M.T. Hoffmann, S. Kreft, G. Pe'er, V. Kati, L. Biber-Freudenberger, D.A. DellaSala, M.M. Vale, P.R. Hobson & N. Selva, 2016. A global map of roadless areas and their conservation status. *Science*, 354: 1423–1427.

Imhoff, M.L., L. Bounoua, T. Ricketts, C. Loucks, R. Harriss & W.T. Lawrence, 2004. Global patterns in human consumption of net primary production. *Nature*, 429: 870–873.

Jaganmohan, M., L.S. Vailshery & H. Nagendra, 2013. Patterns of insect abundance and distribution in urban domestic gardens in Bangalore, India. *Diversity*, 5: 767–778.

Jeevanandam, N. & R.T. Corlett, 2013. Fig wasp dispersal in urban Singapore. *Raffles Bulletin of Zoology*, 61: 343–347.

Jenkinson, F., 1922. Obituary. *The Entomologist's Monthly Magazine*, 58: 20–22.

Johnson, M.T.J., K.A. Thompson & H.S. Saini, 2015. Plant evolution in the urban jungle. *American Journal of Botany*, 102: 1951–1953.

Johnston, M., S. Nail & S. James, 2011. "Natives versus aliens:" the relevance of the debate to urban forest management in Britain. *Proceedings of the conference "Trees, People and the Built Environment,"* Birmingham, UK.

Jones, C.G., J.H. Lawton & M. Shachak, 1994. Organisms as ecosystem engineers. *Oikos*, 69: 373–386.

Jones, D., 2009. Tough start builds urban survivors. *Wildlife Australia*, 46 (3): 43.

Jones, J. & C.M. Francis, 2003. The effects of light characteristics on avian mortality at lighthouses. *Journal of Avian Biology*, 34: 328–333.

Kaplan, S., 2016. These fish evolved to survive the most poisoned places in America. *Washington Post*, December 8, 2016.

Kettlewell, H.B.D., 1955. Selection experiments on industrial melanism in the Lepidoptera. *Heredity*, 9: 323–342.

Kettlewell, H.B.D., 1956. Further selection experiments on industrial melanism in the Lepidoptera. *Heredity*, 10: 287–301.

Kettlewell, H.B.D., 1973. *The Evolution of Melanism*. Oxford: Clarendon Press. 423 pp.

Khanna, P., 2016. *Connectography: Mapping the Future of Global Civilization*. Random House, 496 pp.

Kobori, H. & R.B. Primack, 2003. Conservation for Satoyama, the traditional landscape of Japan. *Arnoldia*, 62 (4): 3–10.

Kohsaka, R., W. Shih, O. Saito & S. Sadohara, 2013. Local assessment of Tokyo: Satoyama and Satoumi—traditional landscapes and management practices in a contemporary urban environment. Pp. 93–105 in: *Urbanization, biodiversity and ecosystem services: Challenges and opportunities*. Springer, the Netherlands.

Kowarik, I., 2011. Novel urban ecosystems, biodiversity, and conservation. *Environmental Pollution*, 159: 1974–1983.

Lampe, U., K. Reinhold & T. Schmoll, 2014. How grasshoppers respond to road noise: developmental plasticity and population differentiation in acoustic signaling. *Functional Ecology*, 28: 660–668.

Lampe, U., T. Schmoll, A. Franzke & K. Reinhold, 2012. Staying tuned: grasshoppers from noisy roadside habitats produce courtship signals with elevated frequency components. *Functional Ecology*, 26: 1348–1354.

Lefebvre, L., 1995. The opening of milk bottles by birds: evidence for accelerating learning rates, but against the wave-of-advance model of cultural transmission. *Behavioral Processes*, 34: 43–53.

Lippens P. & H. van Hengel, 1962. De merel de laatste 150 jaar. Campina.

Le Gros, A., S. Samadi, D. Zuccon, R. Cornette, M.P. Braun, J.C. Senar & P. Clergeau, 2016. Rapid morphological changes, admixture and invasive success in populations of Ring-necked parakeets (*Psittacula krameri*) established in Europe. *Biological Invasions*, 18: 1581–1598.

Littleford-Colquhoun, B.L., C. Clemente, M.J. Whiting, D. Ortiz-Barrientos & C.H. Frère, 2017. Archipelagos of the Anthropocene: rapid and extensive differentiation of native terrestrial vertebrates in a single metropolis. *Molecular Ecology*, 26: 2466–2481.

Lok, A.F.S.L. & T.K. Lee, 2009. Barbets of Singapore Part 2: *Megalaima haemacephala indica* Latham (Coppersmith barbet), Singapore's only native, urban barbet. *Nature in Singapore*, 1: 47–54.

Lok, A.F.S.L., W.F. Ang, B.Y.Q. Ng, T.M. Leong, C.K. Yeo & H.T.W. Tan,

2013. *Native fig species as a keystone resource for the Singapore urban environment*. Raffles Museum of Biodiversity Research, Singapore. 55 pp.

Longcore, T. & C. Rich, 2004. Ecological light pollution. *Frontiers in Ecology and the Environment*, 2: 191–198.

Losos, J.B., 2009. *Lizards in an Evolutionary Tree: Ecology and Adaptive Radiation of Anoles*. Berkeley: University of California Press, 528 pp.

Losos, J.B., K.I. Warheit & T.W. Schoener, 1997. Adaptive differentiation following experimental island colonization in *Anolis* lizards. *Nature*, 387: 70–73.

MacArthur, R.A. & E. O. Wilson, 1967. *The Theory of Island Biogeography*. Princeton: Princeton University Press, 224 pp.

Mahoney, J., 2012. Why wild animals are moving into cities, and what to do about it. *Popular Science*, December 19, 2012.

Majerus, M.E.N., 1998. *Melanism. Evolution in Action*. Oxford, Oxford University Press, 338 pp.

Majerus, M.E.N., 2009. Industrial melanism in the peppered moth, *Biston betularia*: an excellent teaching example of Darwinian evolution in action. *Evo Edu Outreach*, 2: 63–74.

Marlowe, F.W., 2005. Hunter-gatherers and human evolution. *Evolutionary Anthropology*, 14: 54–67.

Marnocha, E., J. Pollinger & T.B. Smith, 2011. Human-induced morphological shifts in an island lizard. *Evolutionary Applications*, 4: 388–396.

Marris, E., 2011. *The Rambunctious Garden: Saving Nature in a Post-Wild World*. New York: Bloomsbury USA, 224 pp.

Martins, C.M. & L.R.L. Simone, 2014. A new species of *Adelopoma* from São Paolo urban park, Brazil (Caenogastropoda, Diplommatinidae). *Journal of Conchology*, 41: 767–773.

Mäser, P., B. Eckelman, R. Vaidyanathan, T. Horie, D.J. Fairbairn, M. Kubo, M. Yamagami, K. Yamaguchi, M. Nishimura, N. Uozumi, W. Robertson, M.R. Sussman & J.I. Schroeder, 2002. Altered shoot/root Na+ distribution and bifurcating salt sensitivity in *Arabidopsis* by genetic disruption of the Na+ transporter AtHKT1. *FEBS letters*, 531: 157–161.

Mayr, E. & W.B. Provine, 1980. *The Evolutionary Synthesis: Perspectives in the Unification of Biology*. Cambridge, MA: Harvard University Press.

McDonnell, M.J. & I. MacGregor-Fors, 2016. The ecological future of cities. *Science*, 352: 936–938.

McGlothlin, J.W., J.M. Jawor, T.J. Greives, J.M. Casto, J.L. Phillips & E.D. Ketterson, 2008. Hormones and honest signals: males with larger ornaments elevate testosterone more when challenged. *Journal of Evolutionary Biology*, 21: 39–48.

McKinney, M.L., 2008. Effects of urbanization on species richness: A review of plants and animals. *Urban Ecosystems*, 11: 161–176.

McNew, S.M., D. Beck, I. Sadler-Riggleman, S.A. Knutie, J.A.H. Koop, D.H. Clayton & M.K. Skinner, 2017. Epigenetic variation between urban and rural populations of Darwin's finches. *BMC Evolutionary Biology*, 17: 183.

Merritt, R.W. & H.D. Newson, 1978. Ecology and management of arthropod populations in recreational lands. Pp. 125–162 in (G.W. Frankie & C.S. Koehler, eds.) *Perspectives in Urban Entomology*. New York: Academic Press.

Miller, P., 2009. Before New York—When Henry Hudson first looked on Manhattan in 1609, what did he see? *National Geographic Magazine*, September 2009 issue.

Misra, T., 2015. East Asia's Massive Urban Growth, in 5 Infographics. (At www.citylab.com)

Misra, T., 2016. Mapping 6,000 Years of Urban Settlements. (At www.citylab.com)

Mockford, E.J. & R.C. Marshall, 2009. Effects of urban noise on song and response behavior in great tits. *Proceedings of the Royal Society of London B*, 276: 2979–2985.

Moeliker, K., 2015. Rotterdamse Natuurvorsers. *Essay Roterodamum*, 2: 1–60.

Moeliker, K., 2016. *De Kloten van de Mus*. Amsterdam: Nieuw Amsterdam.

Møller, A.P., J. Jokimäki, P. Skorka & P. Tryjanowski, 2014. Loss of migration and urbanization in birds: a case study of the blackbird (*Turdus merula*). *Oecologia*, 175: 1019–1027.

Müller, J.C., J. Partecke, B.J. Hatchwell, K.J. Gaston & K.L. Evans, 2013. Candidate gene polymorphisms for behavioral adaptations during urbanization in blackbirds. *Molecular Ecology*, 22: 3629–3637.

Murthy, A.C., T.S. Fristoe & J.R. Burger, 2016. Homogenizing effects of cities on North American winter bird diversity. *Ecosphere*, 7: e01216.

Newitz, A., 2013. *Scatter, Adapt, and Remember; How Humans Will Survive a Mass Extinction*. New York: Doubleday, 305 pp.

Ng, H.H. & H.H. Tan, 2010. An annotated checklist of the non-native freshwater fish species in the reservoirs of Singapore. *Cosmos*, 6: 95–116.

Nielsen, M., 2012. *Reinventing Discovery: The New Era of Networked Science*. Princeton: Princeton University Press, 264 pp.

Nihei, Y., 1995. Variations of behavior of Carrion Crows *Corvus corone* using automobiles as nutcrackers. *Japanese Journal of Ornithology*, 44: 21–35.

Nihei, Y. & H. Higuchi, 2002. When and where did crows learn to use automobiles as nutcrackers? *Tohoku Psychologica Folia*, 60: 93–97.

Nonnekens, A.C, 1961. De Coleoptera van het Amsterdamse bos. *Entomologische Berichten*, 21: 116–128.

Nonnekens, A.C., 1965. De Coleoptera van het Amsterdamse Bos II. *Entomologische Berichten*, 25: 231–233.

Nordt, A. & R. Klenke, 2013. Sleepless in town—drivers of the temporal shift in dawn song in urban European blackbirds. *PLoS One*, 8: e71476.

Nyári, Á., C. Ryall & A.T. Peterson, 2006. Global invasive potential of the house crow *Corvus splendens* based on ecological niche modeling. *Journal of Avian Biology*, 37: 306–311.

Obukhova, N., 2007. Polymorphism and phene geography of the blue rock pigeon in Europe. *Russian Journal of Genetics*, 43: 492–501.

Owen, D.F., 1978. Insect diversity in an English suburban garden. Pp. 13–29 in (G.W. Frankie & C.S. Koehler, eds.) *Perspectives in Urban Entomology*. New York: Academic Press.

Parent, C.E., A. Caccone & K. Petren, 2008. Colonization and diversification of Galápagos terrestrial fauna: a phylogenetic and biogeographical synthesis. *Philosophical Transactions of the Royal Society B*, 363: 3347–3361.

Parker, J., 2016. Myrmecophily in beetles (Coleoptera): evolutionary patterns and biological mechanisms. *Myrmecological News*, 22: 65–108.

Parker, J. & D.A. Grimaldi, 2014. Specialized myrmecophily at the ecological dawn of modern ants. *Current Biology*, 24: 2428–2434.

Parris, K., M. Velik-Lord & J. North, 2009. Frogs call at a higher pitch in traffic noise. *Ecology and Society*, 14: 25.

Partan, S.R., A.G. Fulmer, M.A.M. Gounard & J.E. Redmond, 2010. Multimodal alarm behavior in urban and rural gray squirrels studied by means of observation and a mechanical robot. *Current Zoology*, 56: 313–326.

Partecke, J. & E. Gwinner, 2007. Increased sedentariness in European Blackbirds following urbanization: a consequence of local adaptation? *Ecology*, 88: 882–890.

Partecke, J., I. Schwabl & E. Gwinner, 2006. Stress and the city: urbanization and its effects on the stress physiology in European blackbirds. *Ecology*, 87: 1945–1952.

Partecke, J., T. van 't Hof & E. Gwinner, 2004. Differences in the timing of reproduction between urban and forest European blackbirds (*Turdus merula*): result of phenotypic flexibility or genetic differences? *Proceedings of the Royal Society of London B*: 1995–2001.

Paumgarten, M., 2007. The Mannahatta Project—What did New York look like before we arrived? *The New Yorker*, October 1, 2007 issue.

Peeters, N., 2015. Een botanische misdadiger met een Leidse twist. *De Groene Vinger* (degroenevinger.net), February 25, 2015.

Pennisi, E., 2016. Humans are still evolving—and we can watch it happen. *Science*. https://doi.org/10.1126/science.aaf5727.

Postel, S.L., G.C. Daily & P.R. Ehrlich, 1996. Human appropriation of renewable fresh water. *Science*, 271: 785–788.

Potvin, D.A., K.M. Parris & R.A. Mulder, 2011. Geographically pervasive effects of urban noise on frequency and syllable rate of songs and calls in silvereyes (*Zosterops lateralis*). *Proceedings of the Royal Society of London B*, 278: 2464–2469.

Puntigam, M., J. Braiterman & M. Suzuki, 2010. Biodiversity and new urbanism in Tokyo: The role of the Kanda River. Paper delivered at the International Federation of Landscape Architects World Congress in Suzhou, China.

Reba, M., F. Reitsma & K.C. Seto, 2016. Spatializing 6,000 years of global urbanization from 3700 BC to AD 2000. *Scientific Data*, 3: 160034.

Reid, N.M., D.A. Proestou, B.W. Clark, W.C. Warren, J.K. Colbourne, J.R. Shaw, S.I. Karchner, M.E. Hahn, D. Nacci, M.F. Oleksiak, D.L. Crawford & A. Whitehead, 2016. The genomic landscape of rapid repeated evolutionary adaptation to toxic pollution in wild fish. *Science*, 354: 1305–1308.

Reumer, J., 2014. *Wildlife in Rotterdam; Nature in the City*. Rotterdam Natural History Museum, 158 pp.

Riley, S.P., C. Bromley, R.H. Poppenga, F.A. Uzal, L. Whited & R.M. Sauvajot, 2007. Anticoagulant exposure and notoedric mange in bobcats and mountain lions in urban southern California. *Journal of Wildlife Management*, 71: 1874–1884.

Ripmeester, E.A., M. Mulder & H. Slabbekoorn, 2010. Habitat-dependent acoustic divergence affects playback response in urban and forest populations of the European blackbird. *Behavioral Ecology*, 21: 876–883.

Roth, M. & W.T.L. Chow, 2012. A historical review and assessment of urban heat island research in Singapore. *Singapore Journal of Tropical Geography*, 33: 381–397.

Rothwell, J. & W.A. Lee, 2010. Riverine sediment-associated metal concentrations in the urban tropics: a case study from Singapore. *Geophysical Research Abstracts*, 12: EGU2010-2496.

Rudge, D.W., 2005. Did Kettlewell commit fraud? Re-examining the evidence. *Public Understanding of Science*, 14: 249–268.

Salmon, M.A., P. Marren & B. Harley, 2000. *The Aurelian Legacy: British Butterflies and Their Collectors*. Colchester, Harley Books, 432 pp.

Sanderson, E.W. & M. Brown, 2007. Mannahatta: An ecological first look at the Manhattan landscape prior to Henry Hudson. *Northeastern Naturalist*, 14: 545–570.

Sanderson, E.W., 2009. *Mannahatta: A Natural History of New York City*. New York: Abrams, 352 pp.

Schäfer, M.A., A. Hille & G.B. Uhl, 2001. Geographical patterns of genetic subdivision in the cellar spider *Pholcus phalangioides* (Araneae). *Heredity*, 86: 94–102.

Schilthuizen, M., 2001. *Frogs, Flies, and Dandelions. The Making of Species*. Oxford, Oxford University Press, 245 pp.

Schilthuizen, M., 2008. *The Loom of Life. Unravelling Ecosystems*. Springer, Berlin, 184 pp.

Schilthuizen, M., 2016a. Evolution is happening faster than we thought. *New York Times*, July 23, 2016.

Schilthuizen, M., 2016b. De evolutie ligt op straat. *Bionieuws*, February 13, 2016: 8–9.

Schilthuizen, M., L.P. Santos Pimenta, Y. Lammers, P.J. Steenbergen, M. Flohil, N.G.P. Beveridge, P.T. van Duijn, M.M. Meulblok, N. Sosef, R. van de Ven, R. Werring, K.K. Beentjes, K. Meijer, R.A. Vos, K. Vrieling, B. Gravendeel, Y. Choi, R. Verpoorte, C. Smit & L.W. Beukeboom, 2016. Incorporation of an invasive plant into a native insect herbivore food web. *PeerJ*, 4: e1954.

Schmid, J.A., 1978. Foreword. The urban habitat. Pp. ix-xiii in (G.W. Frankie & C.S. Koehler, eds.) *Perspectives in Urban Entomology*. New York: Academic Press.

Schmidt, D.J.E., R. Pouyat, K. Szlavecz, H. Setälä, D.J. Kotze, I. Yesilonis, S. Cilliers, E. Hornung, M. Dombos & S.A. Yarwood, 2017. Urbanization erodes ectomycorrhizal fungal diversity and may cause microbial communities to converge. *Nature Ecology & Evolution*, 1: 0123.

Schouw, J.F., 1823. *Grundtraek til en almindelig Plantegeographie*. Copenhagen: Gyldendalske Boghandels Forlag, 463 pp.

Secretariat of the Convention on Biological Diversity, 2012. *Cities and Biodiversity Outlook*. Montreal, Canada, 64 pp.

Senar, J.C., M.J. Conroy, J. Quesada & F. Mateos-Gonzalez, 2014. Selection based on the size of the black tie of the great tit may be reversed in urban habitats. *Ecology and Evolution*, 4: 2625–2632.

Serieys, L.E.K., A. Lea, J.P. Pollinger, S.P. Riley & R.K. Wayne, 2015. Disease and freeways drive genetic change in urban bobcat populations. *Evolutionary Applications*, 8: 75–92.

Seto, K. C., B. Güneralp & L.R. Hutyra, 2012. Global forecasts of urban expansion to 2030 and direct impacts on biodiversity and carbon pools. *Proceedings of the National Academy of Sciences*, 109: 16083–16088.

Shapiro, A.M., 2013. Rambunctious Garden: Saving Nature in a Post-Wild World [book review]. *The Quarterly Review of Biology*, 88: 45.

Shochat, E., P.S. Warren, S.H. Faeth, N.E. McIntyre & D. Hope, 2006. From patterns to emerging processes in mechanistic urban ecology. *Trends in Ecology and Evolution*, 21: 186–191.

Shute, P.G., 1951. *Culex molestus. Transactions of the Royal Entomological Society of London*, 102: 380–382.

Silva, C.P., R.D. Sepúlveda & O. Barbosa, 2016. Nonrandom filtering effect on birds: species and guilds response to urbanization. *Ecology and evolution*, 6: 3711–3720.

Silver, K., 2016. The unique mosquito that lives in the London Underground. BBC Earth, March 24, 2016: http://www.bbc.co.uk/earth/story/20160323-the-unique-mosquito-that-lives-in-the-london-underground.

Sin, T.M., H.P. Ang, J. Buurman, A.C. Lee, Y.L. Leong, S.K. Ooia, P. Steinberg & S.L.-M. Teo, 2016. The urban marine environment of Singapore. *Regional Studies in Marine Science*, 8: 331–339.

Sipman, H.J.M., 2009. Tropical urban lichens: observations from Singapore. *Blumea*, 54: 297–299.

Skinner, M.K., 2011. Environmental epigenetic transgenerational inheritance and somatic epigenetic mitotic stability. *Epigenetics*, 6: 838–842.

Slabbekoorn, H., 2013. Songs of the city: noise-dependent spectral plasticity in the acoustic phenotype of urban birds. *Animal Behavior*, 85: 1089e1099.

Slabbekoorn, H. & M. Peet, 2003. Birds sing at a higher pitch in urban noise. *Nature*, 424: 267.

Smith, B.D., 2007. The ultimate ecosystem engineers. *Science*, 315: 1797–1798.

Smith, R.M., K.J. Gaston, P.H. Warren & K. Thompson, 2006a. Urban domestic gardens (IX): Composition and richness of the vascular plant flora, and implications for native biodiversity. *Biological Conservation*, 129: 312–322.

Smith, R.M., P.H. Warren, K. Thompson & K.J. Gaston, 2006b. Urban domestic gardens (VI): environmental correlates of invertebrate species richness. *Biodiversity and Conservation*, 15: 2415–2438.

Soh, M.C.K., N.S. Sodhi, R.K.H. Seoh & B.W. Brook, 2002. Nest site selection of the house crow (*Corvus splendens*), an urban invasive bird species in Singapore and implications for its management. *Landscape and Urban Planning*, 59: 217–226.

Sol, D., A.S. Griffin, I. Bartomeus & H. Boyce, 2011. Exploring or avoiding novel food resources? The novelty conflict in an invasive bird. *PLoS ONE*, 6: e19535.

Soniak, M., 2014. City-dwellers, expect your neighbors to get wilder. *Next City*, October 9, 2014.

SOVON Vogelonderzoek Nederland, 2012. *Atlas van de Nederlandse Broedvogels 1998–2000.—Nederlandse Fauna 5.* Nationaal Natuurhistorisch Museum Naturalis, KNNV Uitgeverij & European Invertebrate Survey-Nederland, Leiden, 584 pp.

Strubbe, D. & E. Matthysen, 2009. Predicting the potential distribution of invasive ring-necked parakeets *Psittacula krameri* in northern Belgium using an ecological niche modeling approach. *Biological Invasions*, 11: 497–513.

Strubbe, D., E. Matthysen & C.H. Graham, 2010. Assessing the potential impact of invasive ring-necked parakeets *Psittacula krameri* on native nuthatches *Sitta europeae* in Belgium. *Journal of Applied Ecology*, 47: 549–557.

Suárez-Rodríguez, M., I. López-Rull & C. Macías Garcia, 2013. Incorporation of cigarette butts into nests reduces nest ectoparasite load in urban birds: new ingredients for an old recipe? *Biology Letters*, 9: 20120931.

Sukopp, H., 2008. On the early history of urban ecology in Europe. Pp. 79–97 in (J. Marzluff *et al.*, eds.) *Urban Ecology: An International Perspective on the Interactions Between Humans and Nature*. Springer, Berlin, 808 pp.

Swaddle, J.P. & R. Lockwood, 2003. Wingtip shape and flight performance in the European Starling *Sturnus vulgaris*. *Ibis*, 145: 457–464.

Swaddle, J.P., C.D. Francis, J.R. Barber, C.B. Cooper, C.C.M. Kyba, D.M. Dominoni, G. Shannon, E. Aschehoug, S.E. Goodwin, A.Y. Kawahara, D. Luther, K. Spoelstra, M. Voss & T. Longcore, 2015. A framework to assess evolutionary responses to anthropogenic light and sound. *Trends in Ecology and Evolution*, 30: 550–560.

Symonds, M.R.E., M.A. Weston, W.F.D. van Dongen, A. Lill, R.W. Robinson & P.-J. Guay, 2016. Time since urbanization but not encephalisation is associated with increased tolerance of human proximity in birds. *Frontiers in Ecology and Evolution*, 4: 117.

Tan, B.C., A. Ng-Chua L.S, A. Chong, C. Lao, M. Tan-Takako, N. Shih-Tung, A. Tay, Y.V. Bing, 2014. The urban pteridophyte flora of Singapore. *Journal of Tropical Biology and Conservation*, 11: 13–26.

Tan, P.Y. & C.Y. Jim, 2017. *Greening Cities: Forms and Functions*. Springer, Singapore, 372 pp.

Tan, H.T.W. & C.K. Yeo, 2009. *The potential of native woody plants for enhancing the urban waterways and water bodies environment in Singapore*. Raffles Museum of Biodiversity Research, Singapore, 28 pp.

Taylor, L.R., R.A. French & I.P. Woiwod, 1978. The Rothamsted insect survey and the urbanization of land in Great Britain. Pp. 31–65 in (G.W. Frankie & C.S. Koehler, eds.) *Perspectives in Urban Entomology*. New York: Academic Press.

Teo, S., K.Y. Chong, Y.F. Chung, B.R. Kurukulasuriya & H.T.W. Tan, 2011. Casual establishment of some cultivated urban plants in Singapore. *Nature in Singapore*, 4: 127–133.

The Data Team, 2016. Bright lights, big cities. Urbanisation and the rise of the megacity. *The Economist*, February 4, 2015.

Thompson, K., 2014. *Where Do Camels Belong? The story and science of invasive species*. London: Profile Books, 272 pp.

Tryjanowski, P., A.P. Møller, F. Morelli, W. Biaduń, T. Brauze, M. Ciach, P. Czechowski, S. Czyż, B. Dulisz, A. Goławski, T. Hetmański, P. Indykiewicz, C. Mitrus, Ł. Myczko, J.J. Nowakowski, M. Polakowski, V. Takacs, D. Wysocki & P. Zduniak, 2016. Urbanization affects neophilia and risk-taking at bird-feeders. *Scientific Reports*, 6: 28575.

Tutt, J.W., 1896. *British Moths*. London: Routledge.

Tüzün, N., L. Op de Beeck & R. Stoks, 2017. Sexual selection reinforces a higher flight endurance in urban damselflies. *Evolutionary Applications*, doi: 10.1111/eva.12485.

Tyler, R.K., K.M. Winchell & L.J. Revell, 2016. Tails of the city: Caudal autotomy in the tropical lizard, *Anolis cristatellus*, in urban and natural areas of Puerto Rico. *Journal of Herpetology*, 50: 435–441.

Uéno, S.-I., 1995. New phreatobiontic beetles (Coleoptera, Phreatodytidae and Dytiscidae from Japan). *Journal of the Speleological Society of Japan*, 21: 1–50.

van Wyhe, J., 2002. *The Complete Work of Charles Darwin Online* (http://darwin-online.org.uk/)

Velguth, P.H. & D.B. White, 1998. Documentation of genetic differences in a volunteer grass, *Poa annua* (annual meadowgrass) under different conditions of golf course turf, and implications for urban landscape plant selection and management. Pp. 613–617 in (J. Breuste, H. Feldmann & O. Uhlmann, eds.) *Urban Ecology*, Springer, Berlin.

Vermeij, G.J., 2012. The limits of adaptation: humans and the predator–prey arms race. *Evolution*, 66: 2007–2014.

Verzijden, M.N., E.A.P. Ripmeester, V.R. Ohms, P. Snelderwaard & H. Slabbekoorn, 2010. Immediate spectral flexibility in singing chiffchaffs during experimental exposure to highway noise. *Journal of Experimental Biology*, 213: 2575–2581.

Vink, J., P. Vollaard & N. de Zwarte, 2017. *Making Urban Nature / Stadsnatuur Maken*. NAI010 Publishers, 320 pp.

Vyas, R., 2012. Current status of Marsh Crocodiles *Crocodylus palustris* (Reptilia: Crocodylidae) in Vishwamitri River, Vadodara City. *Journal of Threatened Taxa*, 4: 3333–3341.

Wang, Y., Q. Huang, S. Lan, Q. Zhang & S. Chen, 2015. Common blackbirds *Turdus merula* use anthropogenic structures as nesting sites in an urbanized landscape. *Current Zoology*, 61: 435–443.

Ward, P., 1968. Origin of the avifauna of urban and suburban Singapore. *The Ibis*, 110: 239–255.

Warren, P.S., M. Katti, M. Erdmann & A. Brazel, 2006. Urban acoustics: it's not just noise. *Animal Behaviour*, 71: 491–502.

Weiner, J., 1995. *The Beak of the Finch: A Story of Evolution in Our Time*. New York: Vintage, 332 pp.

Werf, H. van der, 1982. De bodemfauna van ANS (I). *Natura*, March 1982: 26–30.

White, F.B., 1877. Melanochroism, &c., in Lepidoptera. *Entomologist*, 10: 126–129.

Whitehead, A., B.W. Clark, N.M. Reid, M.E. Hahn & D. Nacci, 2017. When evolution is the solution to pollution: Key principles, and lessons from rapid repeated adaptation of killifish (*Fundulus heteroclitus*) populations. *Evolutionary Applications*, doi:10.1111/eva.12470.

Whitehead, A., W. Pilcher, D. Champlin & D. Nacci, 2011. Common mechanism underlies repeated evolution of extreme pollution tolerance. *Proceedings of the Royal Society of London B*: rspb20110847.

Whitehead, A., D.A. Triant, D. Champlin & D. Nacci, 2010. Comparative transcriptomics implicates mechanisms of evolved pollution tolerance in a killifish population. *Molecular Ecology*, 19: 5186–5203.

Williams, E.H., 2009. Associations of behavioral profiles with social and vocal behavior in the Carolina chickadee (*Poecile carolinensis*). Doctoral Dissertation, University of Tennessee, 193 pp.

Winchell, K.M., R.G. Reynolds, S.R. Prado-Irwin, A.R. Puente-Rolón & L.J. Revell, 2016. Phenotypic shifts in urban areas in the tropical lizard *Anolis cristatellus*. *Evolution*, 70: 1009–1022.

Wittig, R. & U. Becker, 2010. The spontaneous flora around street trees in cities—A striking example for the worldwide homogenization of the flora of urban habitats. *Flora-Morphology, Distribution, Functional Ecology of Plants*, 205: 704–709.

Woodsen, M., 2011. Bad vibrations: the problem of noise pollution. The Cornell Lab of Ornithology, https://www.allaboutbirds.org/bad-vibrations-the-problem-of-noise-pollution/.

Worm, B. & R.T. Paine, 2016. Humans as a hyperkeystone species. *Trends in Ecology and Evolution*, 31: 600–607.

Wright, J.P., C.G. Jones & A.S. Flecker, 2002. An ecosystem engineer, the beaver, increases species richness at the landscape scale. *Oecologia*, 132: 96–101.

Wright, K.M., U. Hellsten, C. Xu, A.L. Jeong, A. Sreedasyam, J.A. Chapman, J. Schmutz, G. Coop, D.S. Rokhsar & J.H. Willis, 2015. Adaptation to heavy-metal contaminated environments proceeds via selection on pre-existing genetic variation. *bioRxiv*, 029900.

Xu, Y., F. Luo, A. Pal, K. Yew-Hoong Gin & M. Reinhard, 2011. Occurrence of emerging organic contaminants in a tropical urban catchment in Singapore. *Chemosphere*, 83: 963–969.

Yang, S. & G. Mountrakis, 2017. Forest dynamics in the U.S. indicate disproportionate attrition in western forests, rural areas and public lands. *PLoS ONE*, 12: e0171383.

Yeh, P.J., 2004. Rapid evolution of a sexually selected trait following population establishment in a novel habitat. *Evolution*, 58: 166–174.

Yong, E., 2012. The catfish that strands itself to kill pigeons. Online at "Not Exactly Rocket Science:" http://blogs.discovermagazine.com/notrocket-science/2012/12/05/the-catfish-that-strands-itself-to-kill-pigeons.

Zala, S.M. & D.J. Penn, 2004. Abnormal behaviors induced by chemical pollution: a review of the evidence and new challenges. *Animal Behavior*, 68: 649–664.

Zerbe, S., U. Maurer, S. Schmitz & H. Sukopp, 2003. Biodiversity in Berlin and its potential for nature conservation. *Landscape and Urban Planning*, 62: 139–148.

ACKNOWLEDGMENTS

First of all, I must thank Peter Tallack and the other people at the Science Factory (Louisa Pritchard and Tisse Takagi), for believing in the book and finding a home for it, and editor Richard Milner at Quercus, and James Meader at Picador, for guiding me through the production process.

My friend and colleague Satoshi Chiba kindly arranged for me to come on a two-months' writing retreat at Tōhoku University in Sendai, Japan, which made it possible to finish the final third of the book in a tranquil and beautiful environment (and which explains the distinctly Japanese flavor of that part of the book). I thank him, his family, his students, and Naito Hiroko for creating such a welcoming atmosphere and for taking my mind off writing by organizing regular caving and other field trips. Suzanne Williams, Ellinor Michel, and Jon Ablett of the Natural History Museum, London, hosted me twice for one week of blissful writing, which effectively took place in the Mollusca department, the NHM restaurant, the Victoria & Albert Museum restaurant, the British Library, Lea Banwell's Bed & Breakfast, and the South Kensington Pret a Manger. Other places that provided, either wittingly or unwittingly, a cushioned writing environment, were, in no particular order, Maliau Basin Studies Center in Malaysian Borneo, Düsseldorf Airport, Hotel Čertousy in Prague, the Ouibus between Paris and Amsterdam, Darko Jesic's Paris apartment, the Groningen University's field center in Schiermonnikoog ("De Herdershut"), Willer highway express bus between Tokyo and Sendai, Flight SQ323 of Singapore Airlines, Ahbam's homestay in Sukau, and the Nexus Karambunai hotel lobby in Kota Kinabalu.

Many scientists and other knowledgeable persons answered questions, proofread bits of text, or supplied me with photos or research materials. I am grateful to: Néstor Alirio, Jacques van Alphen, Florian Altermatt, Garry Bakker, Olga Barbosa, Lin Op de Beeck, Herman Berkhoudt, Pierre-Paul Bitton, Edwin Brosens, Scott Carroll, Jason Chapman, Marion Chatelain, Pierre-Olivier Cheptou, Kayla Coldsnow, Julien Cucherousset, Curt Daehler, Cat Davidson, Luis Fernando De León, Thom van Dooren, Stephanie Doucet, Meghan Duffy, Janko Duinker, Naim Edwards, Clinton Francis,

Max Galka, Kevin Gaston, József Geml, Wouter Halfwerk, Adam Hart, Axel Hochkirch, Bert Hölldobler, Wendy Jesse, Marc Johnson, Masakado Kawata, Gail Kuhnlein, Kate Kuykendall, Ariane Le Gros, Isabel López-Rull, Suzanne MacDonald, Emma Marris, Bennie Meek, Martin Melchers, Osamu Mikami, Erik van Nieukerken, Joe Parker, Jesko Partecke, Carmen Paz, Norbert Peeters, Paloma Plant, Lidy Poot, Alexander Reeuwijk, David Rentz, Jelle Reumer, Ignacio Ribera, Erwin Ripmeester, Milena Salgado-Lynn, Eric Sanderson, Frédéric Santoul, Juan Carlos Senar, Laurel Serieys, Frédérique Soulard, Kamiel Spoelstra, Danica Stark, Monserrat Suárez-Rodriguez, Stephen Sutton, Matt Symonds, Etsuro Takagi, Tan Siong Kiat, Piotr Tryjanowski, Nedim Tüzün, Geerat Vermeij, Oscar Vorst, Gijsbert Werner, Thomas Wesener, Monica Wesseling, Kristin Winchell, John van Wyhe, Bakhtiar Effendi Yahya, and Pamela Yeh.

Several colleagues invested considerably more time by letting me interview them in person or over Skype, or by engaging in lengthy email correspondence. These were Marina Alberti, Jean-Nicolas Audet, Laurence Cook, Karl Evans, Tetsuro Hosaka, Kees Moeliker, Jason Munshi-South, Shinya Numata, Laurel Serieys, Hans Slabbekoorn, Andrew Whitehead, and Niels de Zwarte.

Throughout this book project, friends and colleagues regularly sent me bits of news, social media posts, and scientific papers on urban evolution. Particularly active were Aglaia Bouma, Bronwen Scott, and Rutger Vos, but I also received valuable tips from Thijmen Breeschoten, Tom Van Dooren, Barbara Gravendeel, Marco Roos, and Martin Rücklin. Other sources I would like to acknowledge for their help are the library of Naturalis Biodiversity Center, Wikipedia and its wikipedians, all the readers of my *New York Times* article who contacted me (and especially Michael McGuire and Barbara Waugh), the Rotterdam Natural History Museum, and my students at the Leiden University MSc-course Orientation in Biodiversity and Conservation.

Minoru Chiba and Yawara Takeda took us on a trip through Sendai to find nutcracker-crows. My daughter Fenna Schilthuizen goaded me through Roppongi Hills in Tokyo. Chan Sow-Yan guided me on a voyage through the urban nature of Singapore. Sabine Rietkerk corresponded with me on the whereabouts and fate of the Hoek van Holland house crows.

Auke-Florian Hiemstra's pep-talk gave me the courage to drag myself through the final days of working on the book.

Three people very close to me were willing to proofread the entire manuscript as it matured. These were Aglaia Bouma, Iva Njunjić and Frank van

Rooij, and I am immensely grateful to them for their time, understanding, and clever comments.

Finally, while many people helped me with proofreading and catching errors, I take responsibility for the final content of the text, and the interpretations of research results.

INDEX

ABOUT THE AUTHOR

MENNO SCHILTHUIZEN is a senior research scientist at Naturalis Biodiversity Center in the Netherlands and professor of evolutionary biology at Leiden University. The author of more than 100 papers in scientific literature, he has also written more than 250 stories, columns, and articles for publications including *New Scientist*, *Time*, and *Science*. A frequent guest on radio and television, he is the author of three previous books: *Frogs, Flies and Dandelions* (2001), *The Loom of Life* (2008), and *Nature's Nether Regions* (2014). Together with Iva Njunjić, he runs Taxon Expeditions, which organizes scientific expeditions for lay people.